MULTI-CRITERIA DECISION-MAKING TECHNIQUES IN WASTE MANAGEMENT

A Case Study of India

AAP Research Notes on Optimization and Decision-Making Theories

MULTI-CRITERIA DECISION-MAKING TECHNIQUES IN WASTE MANAGEMENT

A Case Study of India

Suchismita Satapathy, PhD
Debesh Mishra, PhD
Prasenjit Chatterjee, PhD

First edition published 2022

Apple Academic Press Inc.
1265 Goldenrod Circle, NE,
Palm Bay, FL 32905 USA

4164 Lakeshore Road, Burlington,
ON, L7L 1A4 Canada

CRC Press
6000 Broken Sound Parkway NW,
Suite 300, Boca Raton, FL 33487-2742 USA

2 Park Square, Milton Park,
Abingdon, Oxon, OX14 4RN UK

Apple Academic Press exclusively co-publishes with CRC Press, an imprint of Taylor & Francis Group, LLC

Library and Archives Canada Cataloguing in Publication

Title: Multi-criteria decision-making techniques in waste management : a case study of India / Suchismita Satapathy, PhD, Debesh Mishra, PhD, Prasenjit Chatterjee, PhD.
Names: Satapathy, Suchismita, author. | Mishra, Debesh, author. | Chatterjee, Prasenjit, 1982- author.
Description: First edition. | Series statement: AAP research notes on optimization and decision-making theories | Includes bibliographical references and index.
Identifiers: Canadiana (print) 2021020057X | Canadiana (ebook) 20210200669 | ISBN 9781774630136 (hardcover) | ISBN 9781774638750 (softcover) | ISBN 9781003180586 (ebook)
Subjects: LCSH: Refuse disposal industry—India—Case studies. | LCSH: Refuse and refuse disposal—India—Decision making—Case studies. | LCGFT: Case studies.
Classification: LCC HD9975.I53 S28 2022 | DDC 363.72/80954—dc23

Library of Congress Cataloging-in-Publication Data

CIP data on file with US Library of Congress

ISBN: 978-1-77463-013-6 (hbk)
ISBN: 978-1-77463-875-0 (pbk)
ISBN: 978-1-00318-058-6 (ebk)

AAP RESEARCH NOTES ON OPTIMIZATION AND DECISION-MAKING THEORIES

SERIES EDITORS:

Dr. Prasenjit Chatterjee
Department of Mechanical Engineering, MCKV Institute of Engineering, Howrah, West Bengal, India
E-Mail: dr.prasenjitchatterjee6@gmail.com / prasenjit2007@gmail.com

Dr. Dragan Pamucar
University of Defence, Military academy, Department of Logistics, Belgrade, Serbia; E-Mail: dpamucar@gmail.com

Dr. Morteza Yazdani
Department of Business & Management, Universidad Loyola Andalucia, Seville, Spain; E-Mail: morteza_yazdani21@yahoo.com

Dr. Anjali Awasthi
Associate Professor and Graduate Program Director (M.Eng.), Concordia Institute for Information Systems Engineering, Concordia, Canada
E-Mail: anjali.awasthi@concordia.ca

Most real-world search and optimization problems naturally involve multiple criteria as objectives. Different solutions may produce trade-offs (conflicting scenarios) among different objectives. A solution that is better with respect to one objective may be a compromising one for other objectives. This compels one to choose a solution that is optimal with respect to only one objective. Due to such constraints, multi-objective optimization problems (MOPs) are difficult to solve since the objectives usually conflict with each other. It is usually hard to find an optimal solution that satisfies all objectives from the mathematical point of view. In addition, it is quite common that the criteria of real-world MOPs encompass uncertain information, which becomes quite a challenging task for a decision maker to select the criteria. Also, the complexities involved in designing mathematical models increase. Considering, planning, and appropriate decision-making require the use of analytical methods that examine trade-offs; consider multiple scientific, political, economic, ecological, and social dimensions; and reduce possible conflicts in an optimizing framework. Among all these, real-world multi-criteria decision-making (MCDM) problems related to engineering optimizations are categorically important and are quite often encountered with a wide range of applicability.

MCDM problems are basically a fundamental issue in various fields, including applied mathematics, computer science, engineering, management, and operations research. MCDM models provide a useful way for modeling various real-world problems and are extensively used in many different types of systems, including, but not limited to, communications, mechanics, electronics, manufacturing, business management, logistics, supply chain, energy, urban development, waste management, and so forth.

In the aforementioned cases, modeling of multiple criteria problems often becomes more complex if the associated parameters are uncertain and imprecise in nature. Impreciseness or uncertainty exists within the parameters due to imperfect knowledge of information, measurement uncertainty, sampling uncertainty, mathematical modeling uncertainty, etc. Theories like probability theory, fuzzy set theory, type-2 fuzzy set theory, rough set, grey theory, neutrosophic uncertainty theory available in the existing literature deal with such uncertainties. Nevertheless, the uncertain multi-criteria characteristics in such problems are not explored in depth, and a lot can be achieved in this direction. Hence, different mathematical models of real-life multi-criteria optimization problems can be developed on various uncertain frameworks with special emphasis on sustainability, manufacturing, communications, biomedical, electronics, materials, energy, agriculture, environmental engineering, strategic management, flood risk management, supply chain, waste management, transportations, economics, and industrial engineering problems, to name a few.

Coverage & Approach:

The primary endeavor of this series is to introduce and explore contemporary research developments in a variety of rapidly growing decision-making areas. The volumes will deal with the following topics:

- Crisp MCDM models
- Rough set theory in MCDM
- Fuzzy MCDM
- Neutrosophic MCDM models
- Grey set theory

- Mathematical programming in MCDM
- Big data in MCDM
- Soft computing techniques
- Modelling in engineering applications
- Modeling in economic issues
- Waste management
- Agricultural practice
- Material selection
- Renewable energy planning
- Industry 4.0
- Sustainability
- Supply chain management
- Environmental policies
- Manufacturing processes planning
- Transportation and logistics
- Strategic management
- Natural resource management
- Biomedical applications
- Future studies and technology foresight
- MCDM in governance and planning
- MCDM and social issues
- MCDM in flood risk management
- New trends in multi-criteria evaluation
- Multi-criteria analysis in circular economy
- Multi-criteria evaluation for urban and regional planning
- Integrated MCDM approaches for modeling relevant applications and real-life problems

Types of volumes:

This series reports on current trends and advances in optimization and decision-making theories in a wider range of domains for academic and research institutes along with industrial organizations. The series will cover the following types of volumes:

- Authored volumes
- Edited volumes
- Conference proceedings
- Short research (thesis-based) books
- Monographs

Features of the volumes will include recent trends, model extensions, developments, real-time examples, case studies, and applications. The volumes aim to serve as valuable resources for undergraduate, postgraduate and doctoral students, as well as for researchers and professionals working in a wider range of areas.

CURRENT & FORTHCOMING BOOKS IN THE SERIES

Multi-Criteria Decision-Making Techniques in Waste Management: A Case Study of India
Editors: Suchismita Satapathy, Debesh Mishra, and Prasanjit Chatterje

Applications of Artificial Intelligence in Business and Finance: Modern Trends
Editors: Vikas Garg, Shalini Aggarwal, Pooja Tiwari, and Prasenjit Chatterjee

Computational Intelligence for Finance and Marketing in Post-Epidemic VUCA World
Editors: Subhranil Som, Ajay Rana, Prasenjit Chatterjee, Roshani Raut, and Hanaa Hachimi

Multi-Criteria Decision-Making Approaches for Designing Environmental Policies
Editors: Prasenjit Chatterjee, Dragan Pamucar, Morteza Yazdani, and Anjali Awasthi

Statistical Estimation Procedure under Model Based Optimization
Editors: Arnab Bandyopadhyay and Samir Dey

Intelligent Decision Support Systems for Smart City Application
Editors: Loveleen Gaur, Vernika Agarwal, and Prasenjit Chatterjee

ABOUT THE AUTHORS

Suchismita Satapathy, PhD
Associate Professor, School of Mechanical Sciences, KIIT University, Bhubaneswar, India

Suchismita Satapathy, PhD, is an Associate Professor in the School of Mechanical Sciences, KIIT University, Bhubaneswar, India. She has published more than 111 articles in national and international journals and conferences. She has also published many books and e-books for academic and research purpose. She has worked as a special issue editor in many journals, organizes national and international FDP. Her area of interest is production operation management, operation research, acoustics, sustainability, supply chain management, etc. With more than 15 years of teaching and research experience, she has guided many PhD, MTech, and BTech students.

Debesh Mishra, PhD
Researcher, KIIT Deemed to be University, Bhubaneswar, Odisha, India

Debesh Mishra, PhD, has more than 10 years of teaching experience at different engineering colleges at both undergraduate and diploma levels. In addition, he has published more than 20 articles in different reputed journals and international conferences. Also, 10 book chapters are credited to him. His area of interest include ergonomics, occupational health and safety, sustainability, supply chain management, optimization techniques, etc. Dr. Mishra obtained his PhD in Mechanical Engineering from KIIT Deemed to be University (Odisha), India. He completed his MTech with specialization in Industrial Engineering and Management and his BE in Mechanical Engineering.

Prasenjit Chatterjee, PhD
Associate Professor, Mechanical Engineering Department, MCKV Institute of Engineering, India

Prasenjit Chatterjee, PhD, is an Associate Professor and Head of the Mechanical Engineering Department at MCKV Institute of Engineering, India. He has published over 80 research papers in various international journals and peer-reviewed conferences, including the *International Journal of Production Research, Applied Soft Computing, Computers, and Industrial Engineering, Management Decision, Clean Technologies and Environmental Policy, Journal of Cleaner Production, Journal of Natural Fibers, Benchmarking, OPSEARCH, International Journal of Advanced Manufacturing Technology, Materials, and Design,* and *Robotics and Computer Integrated Manufacturing,* to name a few. He has received numerous awards, including Best Track Paper Award, Outstanding Reviewer Award, Best Paper Award, Outstanding Researcher Award, and University Gold Medal. He has been the Guest Editor of several special issues in different Scopus and Emerging Sources Citation Index (Clarivate Analytics) indexed journals. He has authored and edited several books on decision-making approaches, supply chain, and sustainability modeling. He is the Lead Series Editor of AAP Research Notes on Optimization and Decision-Making Theories and Frontiers of Mechanical and Industrial Engineering, Apple Academic Press. Dr. Chatterjee is one of the developers of two multiple-criteria decision-making methods called Measurement of Alternatives and Ranking according to Compromise Solution (MARCOS) and Ranking of Alternatives through functional mapping of criterion sub-intervals into a Single Interval (RAFSI).

CONTENTS

ABBREVIATIONS

3PRLPs	third-party reverse logistics providers
AD	anaerobic-digestion
AHP	analytic hierarchical process
ANP	analytic network process
As	arsenic
BOD	biological-oxygen-demand
BPN	best non-fuzzy performance
CBRI	central building research institute
CCRs	coal combustion residues
Cd	cadmium
CE	combustion-efficiency
CFR	code of federal regulations
CH_4	methane
CIPET	Central Institute of Plastics Engineering and Technology
COD	chemical-oxygen-demand
COPRAS	COmplex PRoportional ASsessment
CPCB	central pollution control board
Cr	chromium
CRRA	California Resource-Recovery-Association
Cu	copper
DEMATEL	decision making trial and evaluation laboratory
EBCS	electronic bonus card system
EoL	end-of-life
EPA	environment protection act
EPR	extended producer responsibility
ETP	effluent treatment plant
EU	European Union
EW	e-wastes
FC	financial constraints
FGD	flue gas desulfurization
FNIS	fuzzy negative ideal solution
FPIS	fuzzy positive ideal solution
GM	genetically modified

HBV	hepatitis-B-virus
HBW	hospital and bio-medical wastes
HIV	human immunodeficiency virus
IAATP	inadequate awareness and training programs
IC&I	institutional, commercial, and industrial
ICT	information and communications technologies
IoT	internet of things
ISFGA	Insufficient Support from Government Agencies
IWMO	improper waste management operational strategy
KMO	Kaiser-Meyer-Olkin
LCA	life-cycle-assessment
LOAF	lack of adequate facilities
LOGPP	lack of green procurement policy
LOSP	lack of segregation practices
LOTMC	lack of top management commitment
MABAC	multi-attributive border approximation-area comparison
MADM	multi-attribute decision making
MAS	multi-agent system
MCDA	multi-criteria decision-analysis
MCGP	multi-choice goal programming
MFA	perfluoro methyl vinyl-ether
MODM	multi-objective decision making
MSW	municipal solid waste
MSWM	municipal solid waste management
Mt	million-tons
N	nitrogen
N_2O	nitrous-oxides
NCBR	National Council for Building Research
NEERI	National Engineering and Environmental Research Institute
NH_3	ammonia
NIS	ideal negative solution
OPC	ordinary Portland cement
OR	odd-ratio
OR	odd-risk
PAYT	pay-as-you-throw
Pb	lead
PFA	perfluoro vinyl-ether
PIS	positive ideal solutions

POPs	persistent organic pollutants
PPD	Plant Protection Department
PPT	plasma-pyrolysis-technology
PVC	poly-vinyl chloride
PVDF	poly-vinylidene-fluoride
PVF	poly-vinyl-fluoride
PW	paper-wastes
QFD	quality function deployment
R&D	research and development
RDF	refuse-derived-fuel
RR	relative-risk
RTCAA	reluctance to change and adoption
SD	sustainable development
SDGs	sustainable development goals
SMS	steel melting shop
SQRT	squares and square root
SSCM	sustainable supply chain management
SWARA	step-wise weight assessment ratio analysis
TOPSIS	technique for order preference by similarity to ideal-solution
TPM	total productive maintenance
URHCW	unauthorized reuse of health care waste
VIKOR	VIse Kriterijumska Optimizacija I Kompromisno Resenje
WEEE	waste electric and electronic equipment
WHO	World Health Organization
WLC	weighted linear combination
WSNs	wireless sensor networks
Zn	zinc
ZW	zero waste

ACKNOWLEDGMENTS

I would like to deeply thank my publisher for giving me an opportunity to publish my content. Our publishers have also helped me a lot by taking precious time to help me carry out my experiments and encouraged me to complete my work accurately within the time I was given to do so. They have been a guiding force with encouragement to perform my work. I would like to thank my husband, Pravudatta Mishra, for encouraging me to complete my book project. My daughter, Meghana, also helped me a lot by giving me relaxation from some of her valuable family time. I would like to express my thanks to my father, Dr. R. B. Satapathy, and my mother, B. P. Mishra, my in-laws, Dr. S. N. Mishra and late Dr. B. M. Mishra, who has inspired me to work on this book. KIIT authority Prof. (Dr.) A. Samant is always an inspiration for me in achieving my goal. My colleagues, friends of KIIT, as well as my students all have equally encouraged me. At last, I would like to thank and owe my contribution before almighty God for invisible help.

—**Suchismita Satapathy, PhD**

I would like to take this opportunity to acknowledge a great many people who have contributed directly or indirectly to compile this book. I am very thankful and grateful to those people who have made this book possible. Further, my sincere gratitude and appreciation to my family members for their wholehearted support and encouragement, without which I would not have been able to complete this professionally enriching and personally satisfying journey.

—**Debesh Mishra, PhD**

PREFACE

This book is a series of pages organized for easy portability and reading. It includes eight chapters. Chapter 1 gives an introduction to this book; Chapter 2 gives a literature review; Chapter 3 discusses barriers to waste management in India by MCDM; Chapter 4 provides a clear-cut idea about a fuzzy-COPRAS approach for effective agricultural waste management in India; Chapters 5 and 6 deal with "making e-waste management successful in India" and "challenges in the management of biomedical wastes in India," respectively; Chapter 7 provides a brief on waste management practices to maintain sustainable supply chain management—a case study on thermal power sector; and finally, Chapter 8 gives the conclusion of this book. Each chapter is well-organized and contains case studies. It is an excellent guide for waste management with a comprehensive insight, which helps both academicians and research.

Waste management and problems related to waste in industries, the agricultural sector, and the domestic sector deal with social-economic and environmental hazards. Waste management mainly deals with environmental parameters that impact industries' social and social-economic life and humans. Because of a lack of proper waste management techniques, many sectors are not ready to perform. Many sectors also lose their productivity. Every industry has to take care of its waste management part properly to maintain sustainability. Without proper industrial waste management, the highly productive industry or sector will be a total failure. Waste management and its methods are typically dealt with by framing and following policies, maintaining environmental regulations, etc. With the use of sensitive internet of things (IoT)-based devices and artificial intelligence (AI) techniques, significant changes often answer waste recycling questions in the industry.

Multi-criteria decision-making (MCDM) entails structuring and planning and solving problems involving multiple criteria. The aim is to support decision-makers facing such issues. Typically, there does not exist a singular optimal solution for such concerns. With the help of MCDM, a far better solution is often found for sustainability and waste management techniques in various industries.

This authored book, *Multi-Criteria Decision-Making Techniques in Waste Management: A Case Study of India,* contains eight chapters that are distinguished but interrelated to each other. The waste management and sustainability issues are clarified, ranked, and analyzed from different case studies, experimental, and data analysis. Each of these extended chapters has undergone full double-blind peer review before being selected for this book. These very informative papers will definitely be helpful to researchers, academicians, and industry persons as well. The waste management problem is widespread and a common problem. All over the world, all developing, developed, and non-developing countries face this problem.

It may be nuclear waste, industrial waste, agricultural waste, domestic waste, or e-waste; it is a problem for all. Many methods, methodologies, techniques have been tried; despite that, it is contaminating humanity. Using biodegradable waste properly will help advance technology [research and development (R&D)] and will also be helpful in creating a new opportunity for business ideas. Recycling non-biodegradable waste will generate a scope of new product development. Hence, the concept of proper waste management cleans our environment and creates new employment opportunities.

The first chapter of the book gives a broad idea of different types of waste and how they are recycled to new useful products worldwide. Notably, the recycling of building material and industrial waste is a significant part of this chapter. The images and the explanation of the procedure in this chapter are new of their kind and will clarify waste-related queries conceptually and distinctly.

The second chapter of the book is a literature survey. Waste administration has been a troublesome subject in developing nations such as India, especially in metropolitan areas. There are countless financial and ecological foundations governing every government, so it is unlikely that a government will acknowledge some exceptional removal strategies as being ideal. The selection of legitimate removal strategies is usually understood as a "multi-standards dynamic issue" or "multi-measures choice examination" approach. An extensive audit of the works completed in setting with waste administration procedures in India has been done in this examination. The ultimate results will help suggest or propose the best decisions for garbage removal and medicine techniques to forestall and shield people from the dangerous effects of waste. This section illustrates the literature surveys based on the category of waste and waste management methods. The essay likewise discusses the computational and MCDM strategies for squandering the board.

In the third chapter, we discuss the "Prioritization of Barriers of Waste Management in India by MCDM." At present, wastes pollute the environment, and this is one of the significant concerns to ensuring a safe environment. Proper transfer and proper treatment of wastes, which are overwhelming, often solve these issues. With these ideas in mind, the message of an unpolluted environment continues to spread throughout the planet. In this way, an effort is made in this exploration to overcome the limitations on the most straightforward possible transfer. Therefore, the absolute best treatment of waste can be accomplished in India.

A comprehensive examination is completed on various sorts of wastes, such as a clinic or bio-clinical wastes (HBW), civil solid wastes (MSW), paper-wastes (PW), and e-wastes (EW), followed by the discussion of some fundamental impediments for every waste. In this manner, "VIKOR analysis" positions all obstacles in reliability with their importance so much that basic methods can take care of squander in India.

The fourth chapter clarifies "A Fuzzy-COPRAS Approach for Effective Agricultural Waste Management in India." Agricultural advancement is often accompanied by squandering from misuse of cultivating techniques and brutal treatment of manufactured materials with concerns for the rural and global conditions. The squander was largely generated because of various forms of agrarian activities. Usually, the squander produced from agriculture development is utilized for various applications through proper waste management. In this investigation, the rural waste management framework in an Indian setting was examined, and a Fuzzy-COPRAS approach was applied to provide a compelling waste management strategy.

In the fifth chapter, titled "Making E-Waste Management Successful in India," we suggest a cross-bred MCDM approach based on Fuzzy-TOPSIS to assess the issues and potential solutions for executing the e-waste management system in India. At that point, dependent on the yield, the "connecting the casual parts exercises with the conventional divisions" positioned first, which is trailed by "tending to on the sheltered removal of e-squander. Both inner and outside to homegrown levels," "forcing perilous e-garbage removal expenses from producers and buyers," "embracing a consultative cycle for a viable e-squander the board plan," "advancing of a mindfulness program on reusing and removal of e-squander," "giving motivators and financed plans to the overall population or businesses related with reusing and removal of e-squander," "limitation and legitimate homegrown systems on bringing in of e-squander," and "drawing in interests in e-squander of the executives segments," separately. The proposed fuzzy-TOPSIS technique is an exact, deliberate, and efficient strategy-creating method that permits strategy makers to come up with steady choices.

The sixth chapter provides details on "Challenges in the Management of Bioclinical Wastes in India." Healthcare waste management has been a significant area of study for two expert scientists. Even though few studies have been conducted for its assessment, this issue needs to be considered in various settings. Furthermore, the shady practices sometimes involved in developing nations, such as India, bring up issues concerning its viability. Using the TOPSIS technique, an attempt was made to investigate the potential obstacles to waste management practices in the Indian medical services areas. The critical boundaries or difficulties acquired were "lacking support from government agencies (ISFGA)" trailed by "insufficient awareness and training programs (IAATP)," "money-related constraints (FC)," and "unapproved reuse of health care waste (URHCW)" that requires suitable contemplation and the executive actions.

The seventh chapter discusses "Waste Management Practices to Maintain Sustainable Supply Chain Management—A Case Study on the Thermal Power Sector." Fundamentally, waste management plants in India are zeroing in on economic issues; yet in addition, pursuing a maintainable policy gracefully ties procedure improvement to achieve their tasks with due thought to social issues. The plants focus on both economic and environmental sustainability issues while improving the quality of their processes. Under the standard of sensible, flexible chain or board systems, the practices are subject to rules, such as the utilization of water, cinders, squanders, and energy with the end goal to make sure that natural, social, and monetary elements remain unaffected. In this part, a top-to-bottom investigation was performed by considering the SSCM practices of the Indian warm force segments. Then the TODIM practice of waste management was conducted among coal-based warm force plants, for example, such as privatized, nationalized, state, and imported coal plants. Moreover, their interrelationships were discovered simultaneously.

In Chapter 8, this book gives diverse novel answers for squander in developing nations, especially in India, even though many of the specialized cycles associated with squander management was transferred from developed countries. There is no pro-familiarity; the costly specialized cycles are not meeting the local needs and cannot be utilized fruitfully. As a result of increasing awareness of the problem and growing pressures on the issue, a new generation of strategy producers has emerged to formulate efficient and effective waste management methods. Thus, increasing awareness of the environment and escalating problems encourages policymakers to design an orderly and effective waste management system. This book accomplishes its objectives by providing and conducting a critical analysis of the applications in MCDM techniques for proper waste management. These MCDM methods are present in a wide range of usage in different decision-making applications and can be more effectively used in the waste management context. It also explains IoT and artificial intelligence techniques to resolve waste management problems.

This book provides insight for policymakers and practitioners regarding waste management policies. Students, researchers, and academics can learn from their concepts and apply these methods and techniques to future research.

CHAPTER 1

INTRODUCTION

1.1 GENERAL BACKGROUND

Waste has been a persistent problem at all times. With the rise in population and economic development, its management has consequently remained a massive predicament because the amount of waste generated increases day by day. Wastes may be defined as "components or objects that are disposed of or are supposed to be disposed of or are required to be disposed of by means of the provisions of national law" (UNEP, 2010). Today, throughout the world, the total amount of waste generation annually, including municipal, industrial, and hazardous wastes, are more than 4 billion tons (Hoornweg and Bhada-Tata, 2012). Also, there are growing global impacts of solid wastes as solid waste management costs may enhance from the present annual $205.4 billion to about $375.5 billion in 2025 (Mavropoulos et al., 2012).

Our day-to-day activities induces a large variety of exceptional wastes springing up from a variety of sources. It is much more expensive to clean up the contaminations in wastes for a longer term in contrast to its prevention at the source. Various countries are dealing with an uphill mission for managing their wastes properly with most efforts to decrease the final volumes and to generate adequate funds for waste management. With the increasing importance of waste management, it becomes a prime requirement for any nation to have the joint responsibility of the citizens, industries, the local governments, as well as the pollution control boards. Usually, collective activities are involved in waste management, such as segregating, collecting, transporting, re-processing, recycling, as well as disposal of diverse categories of wastes. The activities involved in various wastes, as well as waste management, have a varying impacts on energy-consumptions, methane-emissions, carbon-storages, environmental, and human health. For instance, the greenhouse gas emissions are reduced by recycling with the aid of stopping methane emissions from landfills or open-abandons, and in preventing the energy consumption for extracting as well as processing of raw materials. Improper waste management is one of the predominant causes of environmental pollutions. According to the estimation of "The World Health Organization (WHO)," the occurrences of a quarter of the diseases to humankind these days are because of prolonged exposures to environmental pollutions. Although most of the environmental-related diseases are not detectable easily, but these may also be acquired all through childhood stages and gets manifested later in adulthood stage. Moreover, there occurs a social implication of waste owing to the deficiencies in proper management of wastes that disproportionately affects the poorer communities, as these wastes are often dumped on lands adjacent to slums localities leading to health implications. A number of waste-pickers are exposed to hazardous substances on a regular basis as they try to secure the survival of them in addition to their families. The infectious agents from health-care facilities, as well as the other harmful emissions released during the recovery of helpful objects from e-wastes (EW), not only affect the health of waste-pickers, but also contribute to water, land as well as air contaminations.

1.2 WASTE MANAGEMENT

With the growth of public awareness regarding the wastes related problems and with the increase in pressure on urban local bodies and the government for effective and efficient management of wastes, there is a change in the concept of waste as "Materials having no use" to "Resources in the wrong place." In the long run, the cleaning-up of wastes is much more expensive in comparison to the prevention at source. Unless the waste separation at source is practiced, the mixed waste becomes useless as a resource. The society need of thinking of different ways in order to lessen and utilize the wastes to other uses. However, the industry-based waste management is one of the most elaborative waste management procedures covering a huge variety of operations for different streams of wastes as well as a variety of phases of wastes life-cycles.

The term sustainable waste management refers to the management of wastes in such a manner that it is environmentally, socially, and economically feasible and acceptable, that can be achieved by means of strategic-setting up, institutional-capability building, financial-encouragements, techno-economically feasible-technology, public-private joint-ventures and community-participations. There are different approaches for waste management for different categories of wastes as well as for wastes in different geographical localities. At the same time as there are many ways for the classification of wastes, for the purpose of this investigation, the wastes has been classified as domestic, agricultural, bio-medical, industrial (thermal power plants), electronic (e-waste), and municipal solid wastes (MSW), respectively as shown in Figure 1.1. These wastes are mainly generated at some stage in the extraction of raw materials, manufacturing as well as processing of raw materials into intermediate and finished products, the consumption of ultimate products, and other human activities.

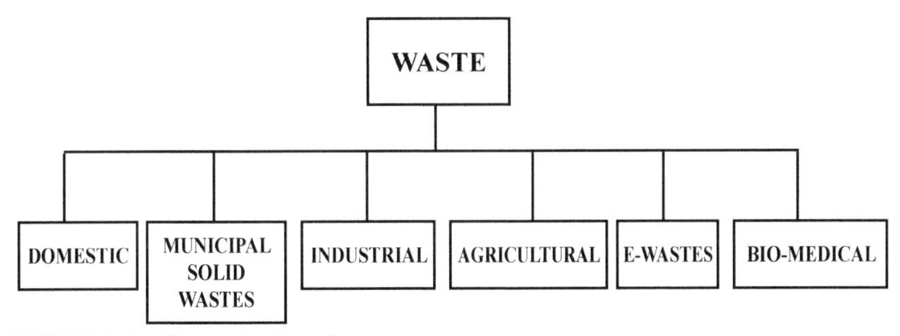

FIGURE 1.1 Categorization of wastes.

The legitimate safeguards regarding the security and protection of the environment has been first provided by India in the world. Different laws regarding waste management in the Indian context are as illustrated in Figure 1.2.

The industrial, commercial as well as institutional sectors may face difficulties in dealing with waste management, which need to take care of a large range of materials, bulky amounts of waste, and activities of many customers, companies from within and outside of the province. None of a specific action will best fit the needs of all the above sectors. However, a strategic waste management planning approach will help to define an effective and rigid solution. A comprehensive strategy can be created by organizations through an integrated waste management planning that will remain flexible in view of the changing social, economic, and environmental conditions. The most efficient and cost-effective technique for waste management in many cases, is not to have to deal with it at all; as a result of which the waste diversion and minimization are often a primary focus for most of the integrated waste management strategies where specific goals as well as targets are well defined. The waste management strategies are as shown in Figure 1.3.

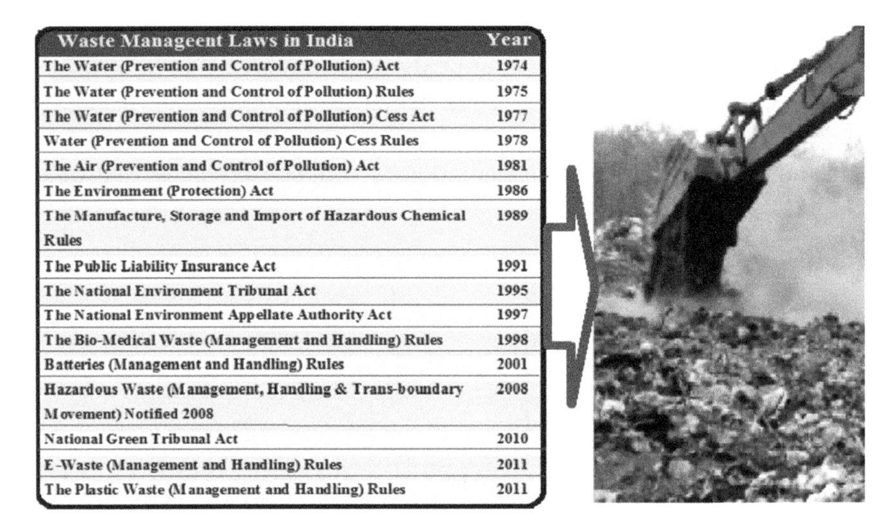

Waste Manageent Laws in India	Year
The Water (Prevention and Control of Pollution) Act	1974
The Water (Prevention and Control of Pollution) Rules	1975
The Water (Prevention and Control of Pollution) Cess Act	1977
Water (Prevention and Control of Pollution) Cess Rules	1978
The Air (Prevention and Control of Pollution) Act	1981
The Environment (Protection) Act	1986
The Manufacture, Storage and Import of Hazardous Chemical Rules	1989
The Public Liability Insurance Act	1991
The National Environment Tribunal Act	1995
The National Environment Appellate Authority Act	1997
The Bio-Medical Waste (Management and Handling) Rules	1998
Batteries (Management and Handling) Rules	2001
Hazardous Waste (Management, Handling & Trans-boundary Movement) Notified 2008	2008
National Green Tribunal Act	2010
E-Waste (Management and Handling) Rules	2011
The Plastic Waste (Management and Handling) Rules	2011

FIGURE 1.2 Different laws concerning waste management in India.

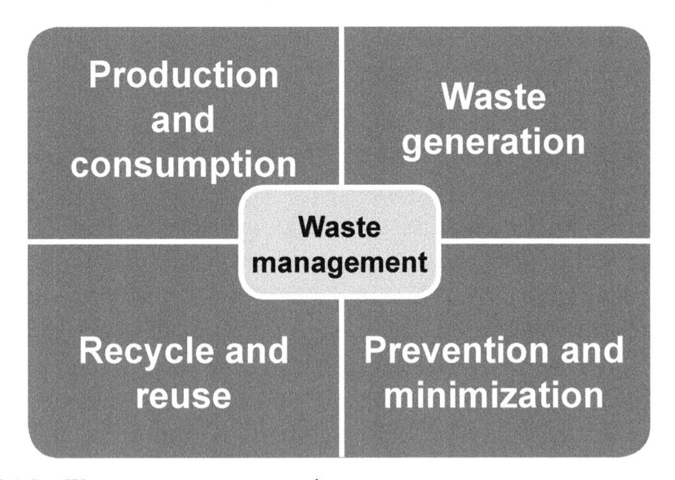

FIGURE 1.3 Waste management strategies.

Proper research enables an essential negotiation to establish the contractual relationship with the waste service providers to ensure that the contract-provisions allow for implementing the waste management strategy successfully. However, a general knowledge of the waste composition as well as volume is required before developing a comprehensive plan which can be typically obtained by conducting studies on waste characterization or waste-audits. Therefore, in the beginning the waste-audit information becomes necessary for planning logistical. Moreover, after implementing the waste management plans, the waste audits are found to be useful to measure the success and progress of the plan, and also for identifying the key areas that require further assessment.

Waste are often classified into five sorts of waste which is all commonly found around the house. These include liquid waste, solid rubbish, organic waste, recyclable rubbish and unsafe waste. Confirm that you simply segregate your waste into these differing types to make sure proper waste removal. Waste is defined as unwanted and unusable materials and is considered a substance which is of no use. Waste that we see in our surroundings is additionally referred to as garbage. Garbage is especially considered as a solid waste that has wastes from our homes (domestic waste), wastes from schools, offices, etc., (municipal wastes) and wastes from industries and factories (Industrial wastes).

Different sources of waste are often broadly classified into four types: industrial, commercial, domestic, and agricultural waste (Figure 1.4).

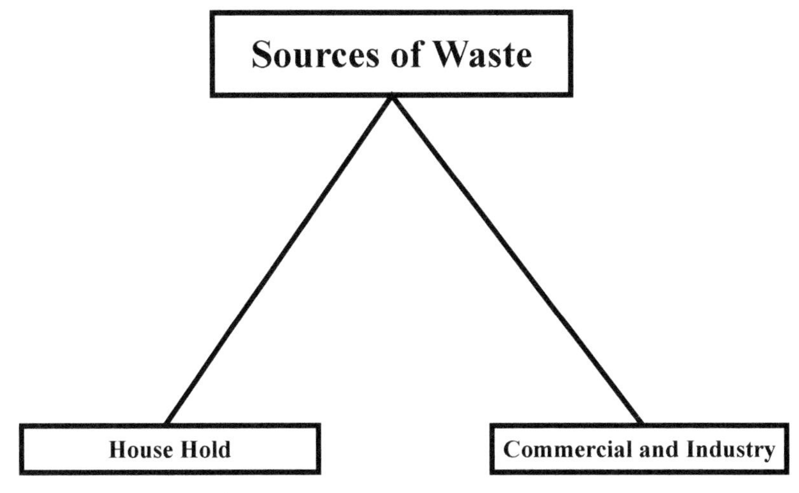

FIGURE 1.4 Sources of waste.

These are the wastes created in factories and industries. Most industries dump their wastes in rivers and seas which cause tons of pollution, e.g., plastic, glass, textile wastes, etc., (Figures 1.5–1.7).

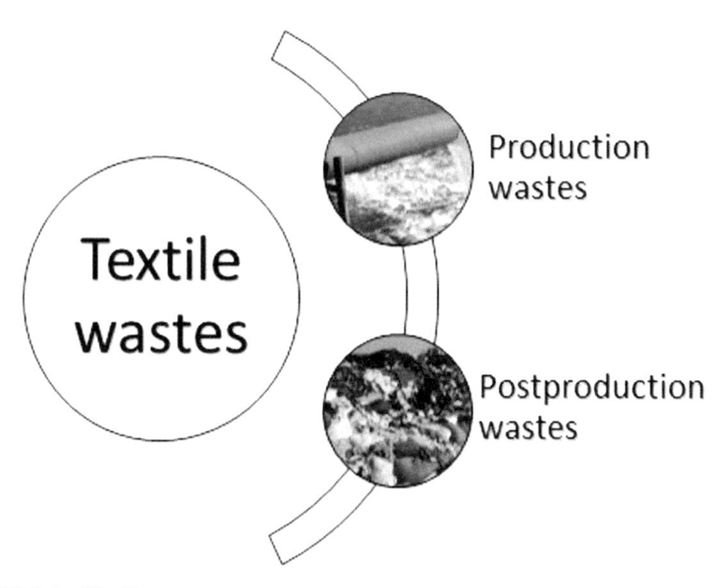

FIGURE 1.5 Textile wastes.

FIGURE 1.6 Plastic waste.

FIGURE 1.7 Glass wastes.

Commercial wastes are produced in shops, schools, colleges, and offices, e.g., plastic, papers, packing-materials, etc. (Figures 1.8 and 1.9).

FIGURE 1.8 Packaging wastes.

FIGURE 1.9 Paper wastes.

The different household wastes which are collected during household activities like cooking, cleaning, etc., are referred to as domestic wastes, e.g., leaves, vegetable-peels, excreta, etc. (Figures 1.10–1.12).

FIGURE 1.10 Leaves waste.

FIGURE 1.11 Vegetable peels.

FIGURE 1.12 Excreta wastes.

Various wastes produced within the agricultural field are referred to as agricultural wastes, e.g., cattle waste, weed, husk, etc. (Figures 1.13–1.15).

FIGURE 1.13 Cattle wastes.

FIGURE 1.14 Weed and husk wastes.

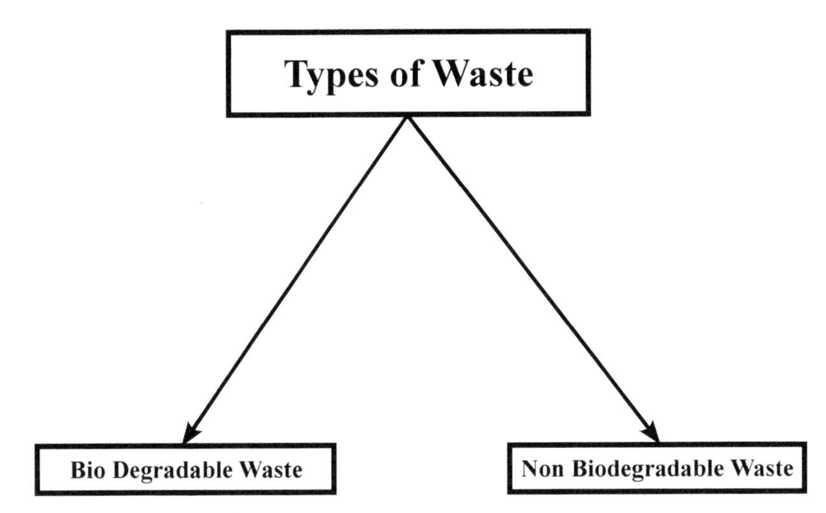

FIGURE 1.15 Types of wastes.

The wastes that come from kitchens include food remains, garden waste, etc. Biodegradable-waste is additionally referred to as moist-wastes that will be composted further to get manures:

1. **Non-Biodegradable Wastes:** These are the wastes which include old newspapers, broken glass pieces, plastics, etc. Non-biodegradable waste is understood as dry waste. Dry wastes are often recycled and may be reused. Non-biodegradable wastes do not decompose by themselves and hence are major pollutants.
2. **Recycling of Wastes:** This is extremely important as this process helps in processing wastes or used products into useful or new products. Recycling helps in controlling air, water, and land pollution. It also uses less energy. There are a variety of things which will be recycled like papers, plastics, glasses, etc. Recycling helps in conserving natural resources and also helps in conserving energy. Recycling helps in protecting the environment as it reduces air, water, and soil pollutions.
3. **Decomposition of Biodegradable Wastes:** Biodegradable-wastes are often decomposed and converted into organic-matters with the assistance of various processes.
4. **Composting:** This is the tactic during which wastes are often decomposed and converted into organic-matters by burying them

within the compost-pits. The wastes are composed by the action of bacteria as well as fungal actions.

5. **Vermicomposting:** This method involves decomposition of organic-matters into fertile-manures with the assistance of red-worms. These manures are understood as vermicompost.
6. **Chemical-Wastes:** These wastes are made up of harmful-chemicals which are mostly produced in large-factories. Chemical-wastes may or may not be hazardous. The chemical-wastes which are hazardous include in the form of solid, liquid or gaseous, and can show hazardous characteristics like corrosivity, toxicity, ignitability, and reactivity.

1.2.1 GENERATION OF WASTES

Usually, waste is generated during our day-to-day activities in household work in industry, in transportation, due to overproduction, during processing, etc.

When coal is transported in trucks, bins or by railway, it is the most polluting waste that not only hazardous for human health, it pollutes air, water, and land. Textile wastes are generated during production as well as after post-production (Figure 1.16).

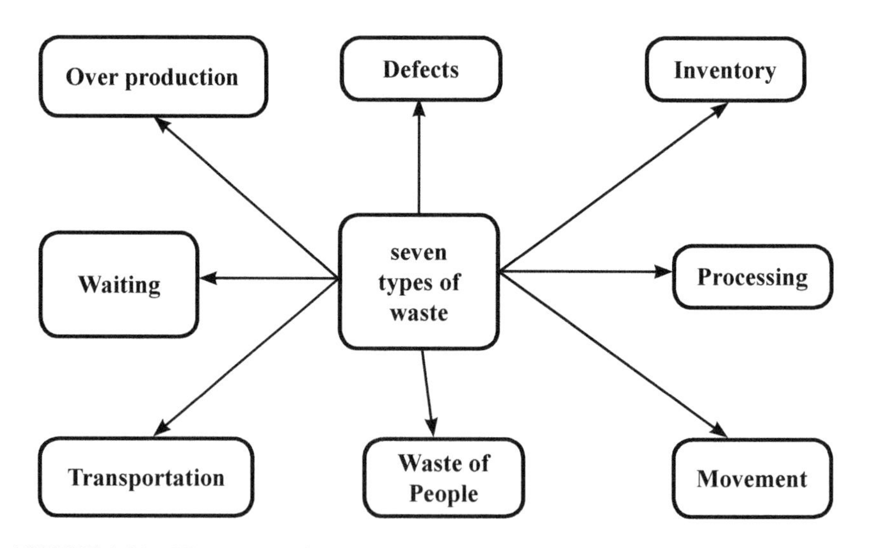

FIGURE 1.16 Waste generation.

1.2.2 WASTE MANAGEMENT METHODS

Waste management methods are (Figure 1.17):

- Source-reduction;
- Reusing;
- Animal-feedings;
- Composting;
- Recycling;
- Fermentations;
- Landfills;
- Incineration and land-applications.

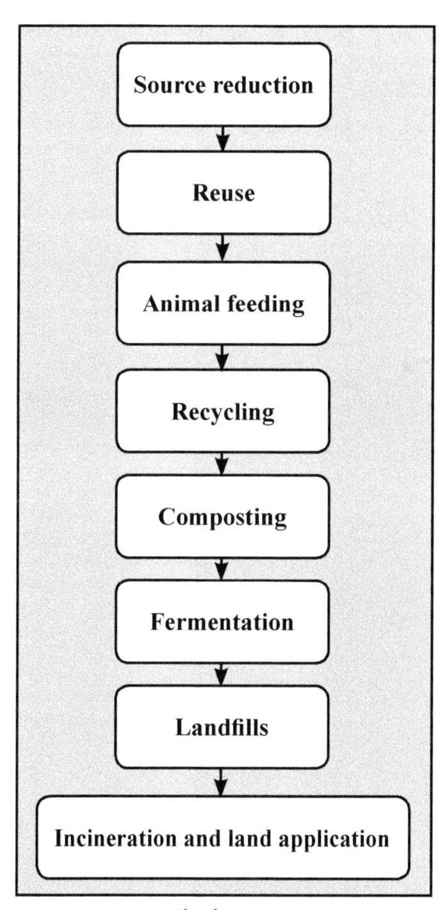

FIGURE 1.17 Waste management methods.

Figure 1.18 illustrates the solid waste management strategy.

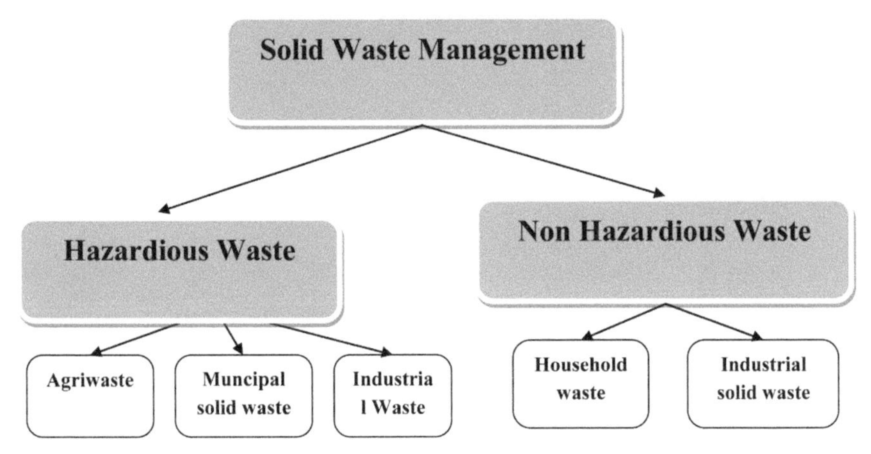

FIGURE 1.18 Solid waste management.

Figure 1.19 illustrates the liquid waste management strategy.

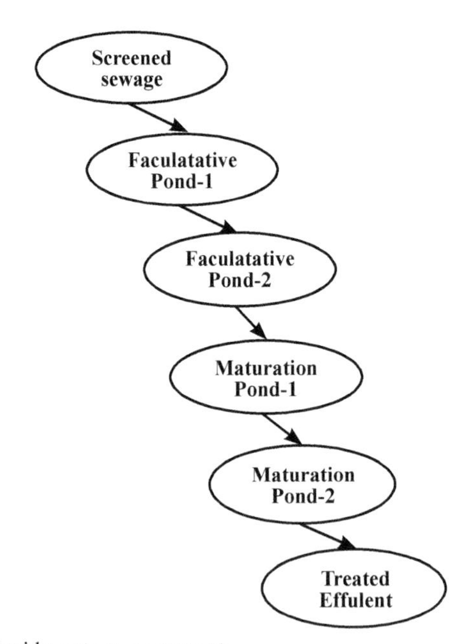

FIGURE 1.19 Liquid waste management.

Figure 1.20 illustrates the hazardous waste management strategy.

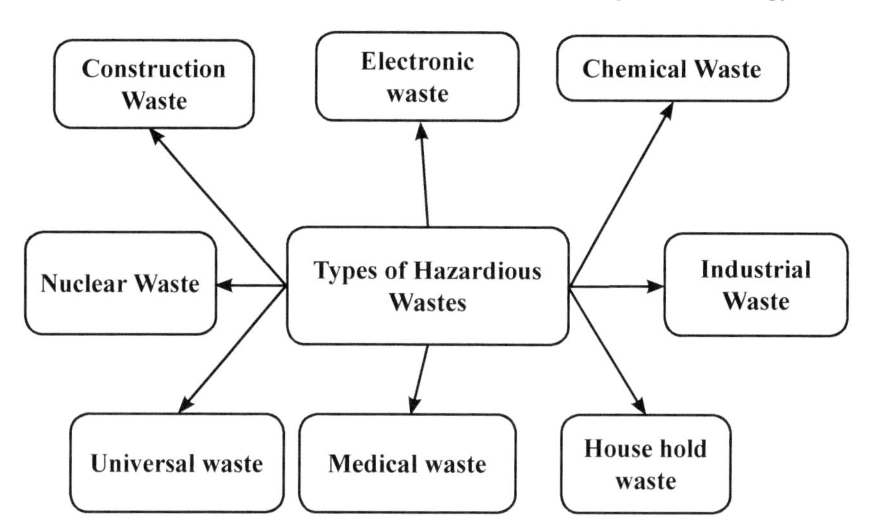

FIGURE 1.20 Hazardous waste types and management.

Different waste management rules are illustrated in the following sections:

➢ **The Hazardous Wastes (Management, Handling, and Trans-boundary Movement) Rules, 2008:** A complex set of rules are incorporated in the management of hazardous waste which are combined together to form the legal-regime. The responsibility for the safer as well as environmentally sound handling of environmental wastes are established by the rules for any "occupier" of hazardous wastes:

- A person under his charge of any plant or factory that produces hazardous wastes or holding of hazardous wastes, is an "occupier." "Recycling" refer to reclamations or re-processing of hazardous wastes in a manner that is environmentally sound for the original and/or other purposes. "Reuse" refer to the use of a hazardous wastes for original and/or other purposes.
- Hazardous-wastes hold by any "occupier" must be sold or sent to recyclers or re-processors having authorizations to dispose it off in proper-manners. In addition, for handling of hazardous-wastes, the responsible holders or persons must

take all necessary steps to prevent from polluting-substances and unsafe effects on human-being health as well as the environment. Besides, necessary training and equipment need to be provided to persons working on sites that are exposed to hazardous-wastes.

- Persons must obtain authorizations from the State Pollution Board, who are actively engaged in the generation, treatments, processing, packaging, storage, transportation, uses, collections, and destructions, conversions, offering for sales and any occupiers.
- The hazardous wastes may be stored for a period of up to 90 days by the occupiers, recyclers, reprocessors, operators, and reusers of facilities, and records of storages, sales, transferring, recycling as well as reprocessing of such wastes should be made available for inspections.
- Persons who wish to recycle or re-process any wastes require of making applications by submission of suitable forms as found in the rules, and the necessary documents include the following:
  - o Consent of establishment that should be granted by the State Pollution Control Board.
  - o Registration certificate issued by the District Industries Centre or any other governmental agency.
  - o Installed-capacity proof of plants and machineries issued by the District Industries Centre or any other governmental agency.
  - o For renewable, compliance-certificate of effluents, emission-standards, treatment, and disposals of hazardous wastes as need to be obtained applicable, from the State Pollution Control Board and the relevant Zonal Office of Central Pollution Control Board (CPCB).
- Occupiers of waste may only transfer or sell to a recyclers with valid-registration from the CPCB.
- Using of hazardous wastes as a resource of energy recovery should be carried out by the units after getting approvals from the CPCB.
- Importing of hazardous wastes are legally prohibited in India, while this may be allowed for the purpose of recovery or recycling or reusing.

- Exporting of hazardous wastes may be allowed with prior informed-consents from the importing country to an actual user of the wastes or disposal facility operators, in order to ensure sound management of the hazardous wastes environmentally.
- Any person intended in importing or transiting hazardous wastes should apply by proper completion of forms that must be submitted to the "Central Government" of the proposed-import along with the "Prior Informed Consent." At the same time, a copy needs to be sent to the concerned "State Pollution Control Board" to enable them in sending their necessary-comments, if any, to the "Ministry of Environment and Forests."
- The examination and grant of the application by "The Ministry of Environment and Forests" is based on the conditions that the importer is having environmentally-sound facilities for recycling, recovery or reuse; adequate facilities as well as arrangements have been generated for the treatments and disposals of wastes; and valid-registration from the "CPCB" and proof of use, if required, have been obtained. Moreover, the importer should also inform regarding the date and the arrival time of the consignment of hazardous wastes to the concerned "State Pollution Control Board" and the "CPCB," 10 days in advance.
- Persons intended in operating some "facility" involved for handling, collections, receptions, treatments, storages, and recycling, reuse, recovery, and disposals of hazardous wastes, need to design and set-up the treatment, storage, and disposal facilities in compliances with the technical-guidelines as issued by the "CPCB." Further, approval from the "State Pollution Control Board" for designing and layout is necessary in this regard from time to time. For the safer as well as environmentally sound operation of facilities and any closures, the operators of the treatments, storages, and disposals facilities shall be held responsible, and in this regard, the issued guidelines by the "CPCB" must be followed.
- All the hazardous wastes are required to be well ensured for packaging and labeling by the occupiers or operators of a 'treatment, storage, and disposal facility' or recyclers, based on the composition in a suitable manner for safe storage, handling, and transport as per the "CPCB" guidelines. There should be clear visibility and it should withstand the physical-conditions and climatic-factors.

- If any occupier intends to transfer the hazardous wastes, then necessary forms need to be filled up and the transporters must be provided with the relevant information regarding the hazardous-natures of the wastes along with the measures to be taken, in case of emergencies. The hazardous wastes should be marked with containers in conformity to the forms.
- "No Objection Certificate" must be obtained from the "State Pollution Control Board" of the States, in case of transporting of hazardous wastes for final-disposal to any facility for treatments, storages, and disposals into some other state than its original generated state. The concerned "State Pollution Control Boards" should be notified by the occupier before handling over the hazardous wastes to the transporter, in such cases.
- The occupiers, importers or operators of facilities are held responsible for any damages that are caused to the environment or third-party due to improper-handling of the hazardous wastes or disposing of the hazardous wastes, and financial-penalties as levied by the "State Pollution Control Board" will be applicable to them in such cases with prior-approval of the "CPCB."

Table 1.1 signifies the color-codes and the relevant-authorities for intra-boundary movements of wastes.

TABLE 1.1 Color-Code and Relevant-Authority for Intra-Boundary Movements of Wastes

Copy-Number	Color-Code	Purposes
Copy-1	White	To be forwarded to the destinated "State Pollution Control Board or Committee." Further, forwarding of a copy is required to the "State Pollution Control Board" of any state of proposed transitions.
Copy-2	Yellow	The occupier must carry after signing from the transporters and the remaining four numbers of copies need to be carried by the transporters.
Copy-3	Pink	The operators must retain the facility after signatures.
Copy-4	Orange	The operators must return the facility or recycler to the transporters after wastes acceptance.
Copy-5	Green	After treatment and disposal of wastes, the operators must return the facility to the "State Pollution Control Board/Committee."
Copy-6	Blue	After treatment and disposal of hazardous wastes, the operators must return the facility to the occupiers.

- **Biomedical Waste (Management and Handling) Rules, 1998:** A proper disposal of biomedical wastes is the key aim of this regulation that has been defined as any wastes that is generated during the diagnosis, treatments or immunizations of human beings or animals or in research-activities pertaining thereto, or testing or production of biological:
  - An occupier refer to someone in relation to any institution that is generating biomedical wastes such as hospitals, nursing homes, clinic dispensaries, veterinary institutions, animal houses, pathology-laboratories, blood banks. The occupier must be ensured of any treatment of wastes that is treated in a manner not harmful to human health or the environment.
  - The occupiers' responsibility is in setting-up the requisite-shall in conformity to the essential biomedical wastes treatment-facilities or in ensuring the necessary treatments of wastes at a common-facility for wastes treatment or any other facility of wastes treatment.
  - The mixing of biomedical wastes with other wastes are prohibited in the rules. It is established that bio-medical wastes should be segregated into bags or containers at the generation-points prior to storage, transportations, treatments as well as disposals. Also, the containers should be labeled in accordance with Schedule III in the rules.
  - No untreated bio-medical wastes should be kept in storages beyond 48 hours period. For any reason, if it becomes necessary for storing the wastes beyond such period, then permission must be taken from the authority by the authorized person as prescribed by the State Government and Union Territory. Adequate measures must be taken in order to ensure that the wastes do not affect adversely the human-health as well as the environment.
  - Application to the established authority by the State-Government and Union Territory is required for the grant of authorization to every occupier of any institution involved in the generation, collection, receipt, storage, transportation, treatment, disposal, and handling of biomedical wastes except dispensaries, clinics, pathological-laboratories, and blood-banks providing services or treatments to lesser than 1000

patients-per-month. Furthermore, every occupier must submit a report by January 31 of every year, that must be included with information about the bio-medical wastes' quantities and categories handled during the previous year.

- Records need to be maintained related to the generation, collection, receipt, storage, transportations, treatments, disposals, and any form of bio-medical wastes handling in accordance to the rules subjected to inspections and verifications.
- The authorized person should report any occurrences of accidents at any institutions or facilities to the prescribed authority.

Table 1.2 illustrates the disposal methods of bio-medical wastes as required by the rules.

➢ **The E-Waste (pManagement and Handling Rules, 2010):** The e-waste rules concerning electronic-wastes is intended to tackle the problematic-situations of disposing of electronic wastes in India, which refers to the generated wastes in India in addition to the larger volumes of wastes illegally imported for disposals into India:

- Electronic wastes are electrical and electronic equipment intended to be discarded, including but is not confined to the listed equipment in the rules, scraps, and rejects from the manufacturing processes.
- A "producer" is any person manufacturing and offering to sell of electrical and electronic equipment under own-brands; or offering to sell any assembled electrical and electronic equipment which are produced by some other suppliers or manufacturers under his own brand; or offering to sell any imported electrical and electronic equipment.
- The producers should collect the EW generated during the electrical and electronic equipment manufacture and to ensure that these are sent for recycling or disposals. The same also applies to wastes collected from producers, electronic wastes generated by-products which have reached their "end-of-life (EoL)." The rules mention the principle of "Extended-Producer responsibility," which means that the extension of a producer's responsibility is not only throughout the products' lifetime, but also to the proper product-disposals.

TABLE 1.2 Disposal Methods of Bio-Medical Wastes as Required by the Rules

Category	Waste-Category (Sources)	Treatment and Disposals
I.	Human-anatomical wastes (human-tissues, organs, body-parts)	Incinerations or deep-burials
II.	Animal-Wastes (Animal-tissues, organs, body-parts carcasses, bleeding-parts, fluids, bloods, and experimental animals utilized in researches, wastes generated by veterinary-hospitals, colleges, discharges from hospitals and animal-houses)	Incinerations or deep-burials
III.	Microbiology and Biotechnology Wastes (Wastes from laboratory-cultures, specimens or storages of live microorganisms or attenuated-vaccines, animal, and human cell-cultures used in researches and infectious-agents from researches and industrial-laboratories, wastes from manufacture of toxins, biologicals, dishes, and the devices utilized for cultures transfer)	Local-autoclaving/ incineration/microwaving
IV.	Waste-sharps (Syringes, needles, blades, scalpels, glasses, etc., causing punctures and cuts, and including both used as well as unused sharps)	Chemical-treatments/ autoclaving/microwaving and mutilations
V.	Discarded-medicines and cytotoxic-drugs (outdated, contaminated, and discarded medicines).	Destruction/incineration and destroyed drugs in landfills
VI.	Contaminated-materials (contaminated-items with bloods and body-fluids that include cottons, dressings, soiled plaster-casts, beddings, linens, and other contaminated materials with bloods).	Incineration/autoclaving/ microwaving
VII.	Disposable-wastes (the generated wastes from disposable solid-items other than waste-sharps such as catheters, tubing, intravenous sets, etc.)	Disinfection by chemical-treatments/autoclaving/ microwaving and mutilations/shredding

- For the generated electronic wastes, collection centers are needed to be set-up. Also, the producers are required for financing or organizing a system to meet the costs incurred for an environmentally-sound e-waste management that are generated from the "EoL" of the products. The financing should be transparent.
- Important aspects of placing the obligation on the producers are: creating awareness through publications, posters, and advertisements regarding the hazardous-constituents in electrical wastes

in addition to the information related to the hazards of improper-handling of the wastes, and instructions to handle the equipment after its' uses.

- Producers are required to obtain authorization from concerned "State Pollution Control Board."
- Producers need to apply within three months of the entry into force of the rules, which must be done by filling required forms and submitting to the "State Pollution Control Board" or the "Pollution Control Committee." The persons who are authorized under the provision of the "Hazardous Wastes (Management, Handling, and Transboundary Movements) Rules, 2008," are not required to make any authorization application till the expiry-period of such authorizations.
- The persons who are authorized under these rules should maintain the records of handled EW by them in required forms, and accordingly, need to prepare as well as submit to the "State Pollution Control Board" or "Pollution Control Committee," with an annual return containing the details on or before 30th day of June following to the related financial-year.
- Applications for renewal of authorizations should be made at least two-months (60-days) before its expiry by appropriate forms to the "State Pollution Control Board" or "Pollution Control Committee." Violating any conditions in the preceding grant of authorizations will disallow the grant of authorizations.

The hazardous-substances that require a permit for "recycling and reprocessing" is as listed below:

- Brass-dross.
- Copper-dross.
- Copper-oxide.
- Copper-reverts, cakes, and residues.
- 'Insulated copper-wire scraps' or 'copper-with-PVC sheathing' including 'IRSI-code' material namely "Druid."
- Copper-cables filled with jelly.
- Lead-acid battery-plates and other lead-scraps/ashes/residues that are not covered under the "Batteries (Management and Handling Rules, 2001)," i.e., 'Lead battery-plates' covered by ISRI Code-word "Rails," 'Battery-lugs' covered by ISRI Code-word "Rakes."
- Paints and ink-sludges/residues.

- Copper-processing generated slags for further 'processing or refining.'
- Spent cleared-metal catalyst containing coppers.
- Spent cleared-metal catalysts containing zinc.
- Spent-catalyst containing "Nickel, Zinc, Cadmium, Copper, Vanadium, Arsenic, and Cobalt."
- Used oils and waste-oils.
- Waste electrical and electronic assembles' components including accumulators as well as other batteries, cathode ray tubes' activated glass-cullets, mercury-switches, and other activated-glasses and PCB capacitors, or any other contaminated components like cadmium, lead, mercury, polychlorinated-biphenyl exhibiting or to an extent of exhibiting hazardous-characteristics.
- Waste-copper and copper-alloys in dispersible-forms.
- Zinc-ashes and residues including zinc-alloy residues in dispersible-forms.
- Zinc-ashes/skimming that arise from die-casting and galvanizing processes.
- Zinc-ashes/skimming/other zinc-bearing wastes that arise from smelting and refining.
- Zinc-dross-bottom dross.
- Zinc-dross-hot dip galvanizers.

1.2.3 THE BASEL-CONVENTION

The most widespread global environmental-agreement regarding harmful and other wastes has been regarded to be "The Basel-Convention" on the Transboundary Movements Control for the hazardous wastes in addition to their disposals. It was enforced on 5 May 1992 with the sign of 173 countries on 22 March 1989, and was mainly created for the prevention of the economic and motivated hazardous wastes dumping from more affluent to lesser affluent countries. Moreover, on the "10th Anniversary of the Basel-Convention in December 1999" on hazardous wastes, the Government-Ministers assembled in Basel (Switzerland) and adopted an environmentally sound management through a declaration on hazardous wastes, which was emphasized on the urgent need for hazardous wastes generation reduction-both in terms of hazardousness as well as quantity. The specific activities under the declaration in the convention (UNEP News Release, 1999) were as follows:

- Active endorsement of cleaner technologies along with production processes;
- Reduction in the movement of hazardous as well as other wastes;
- Prevention and control of illegal-traffics;
- The institutional as well as technical capabilities improvement through appropriate technologies;
- Developing of regional and sub-regional centers for the purpose of technology transfer and training;
- Enhancing the information-exchanges, educations, and growing-awareness in all sectors of society.

Bio-medical wastes refer to any generated wastes during diagnosis, treatments or immunizations of human beings or animals, or during research-activities, or in the manufacture or test of biologicals (Government of India, 1995). Based on the categories of bio-medical wastes, ten categories ranging between category code no. I to 10 (Government of India, 1995), are as follows:

- Human-anatomical wastes that include tissues, organs, body parts.
- Animal-wastes that include animals utilized for research activities and the wastes originating from veterinary-hospitals and animal-houses.
- Micro-biological and biotechnology wastes that include wastes from lab-cultures, microorganisms' specimens, attenuated or live vaccines, biological productions related wastes, etc.
- Waste-sharps that include used and unused needles, lancets, syringes, blades, scalpels, glasses, etc.
- Discarded-medicines and cytotoxic-drugs.
- Soiled-wastes that include contaminated-items with body and blood fluids, cotton-dressings, linens, plaster-casts, etc.
- Solid-wastes that include the generated wastes from disposable-items other than waste-sharps such as catheters, tubing, etc.
- Liquid-wastes that include the generated wastes from cleaning, washing, house-keeping, and disinfection-activities inclusive of lab-activities.
- Incineration-ashes from any bio-medical wastes' incineration.
- Chemical-wastes that include the chemicals used in making of biologicals as well as disinfection.

The different categories of wastes as derived from the "Basel Convention" is illustrated in Table 1.3.

TABLE 1.3 Wastes Categories Derived from the "Basel Convention"

Basel-Number	Description
A-1010	Metal-wastes and wastes comprising of alloys of antimony, lead, cadmium, tellurium.
A-1020	Wastes excluding metal-wastes in massive-forms, with constituents or as contaminants of the following: Lead or Lead-compounds, antimony or antimony-compounds, cadmium or cadmium-compounds, tellurium or tellurium-compounds.
A-1040	Wastes with constituent of metal-carbonyls
A-1050	Galvanic-sludges
A-1060	Waste-liquors from the metals-picking
A-1070	Leaching-residues from dusts and sludges, zinc-processing
A-1080	Waste zinc-residues containing concentrations of cadmium and lead in sufficient quantity to exhibit hazardous-characteristics
A-1090	Ashes generated from the incineration of insulated copper-wires
A1100	Dusts and residues from 'gas clearing-systems of copper-smelters'
A-1110	Spent electrolytic-solutions from 'Copper electro-refining and electrowinning-operations'
A-1120	Waste-sludges excluding anode-slimes from electrolytic-purification systems in copper-electro-re-refining and electro-winning-programs
A-1130	Spent etching-solutions containing dissolved-papers
A-1150	Precious metal-ashes from incineration of 'printed circuit boards
A-1160	Waste of complete or crushed 'lead-acid batteries'
A-1170	Unsorted waste-batteries
A-1180	Waste electrical and electronic assembles' components including accumulators as well as other batteries, cathode ray tubes' activated glass-cullets, mercury-switches, and other activated-glasses and PCB capacitors, or any other contaminated components like cadmium, lead, mercury, polychlorinated-biphenyl exhibiting or to an extent of exhibiting hazardous-characteristics.
A-2010	'Activated glass-cullets' from cathode ray tubes and other activated-glasses
A-2030	Waste-catalysts excluding waste specification in 'List B of Schedule 3' in the rules
A-3010	Waste from the production or processing of Petroleum coke and Bitumen

TABLE 1.3 *(Continued)*

Basel-Number	Description
A-3020	Waste mineral oils unfit for their original intended use
A-3050	Waste from productions, formulations, and uses of resins, latex-plasticizers or adhesives/glues
A-3070	Phenol wastes, phenol-compounds inclusive of chloro-phenol in the form of sludges or liquids
A-3080	Waste-ethers
A-3120	Fluffs: Lighter fraction from 'shredding'
A-3130	Organic phosphorous-compounds' wastes
A-3140	Non-halogenated organic-solvents' wastes
A-3160	Wastes of halogenated or non-halogenated nonaqueous-distillation resides that arise from organic-solvent recovery operations
A-3170	Wastes from the production of aliphatic halogenated-hydrocarbons such as dichloroethane, chloromethanes, vinyl chlorides, vinylidene-chlorides, allyl-chlorides, and epichlorohydrin.
A-4010	Wastes in producing, preparing, and using of pharmaceutical-products
A-4040	Wastes from the manufacture-formulations and utilization of wood-preserving chemicals
A-4070	Wastes in producing, formulating, and using of dyes, inks, pigments, paints, varnishes, and lacquers.
A-4080	Wastes of explosive-materials
A-4090	Wastes of acidic/basic solutions
A-4110	Wastes from 'industrial control devices utilized in industrial off-gases cleaning activities
A-4120	Wastes comprising of or contaminated with peroxides
A-4130	Wastes of packaging and containers that contain any of the constituents as mentioned in 'Schedule 2' of the rules to the specified concentration limits-extent.
A-4140	Wastes comprising of outdated chemicals or that contain off-specification as mentioned in 'Schedule 2' of the rules.
A-4150	Waste chemical-substances coming up from teaching and R&D activities which are identified to have effects on human health and/or the environment and/or are new.
A-4160	Spent-activities carbon

The requirement of different permissions by various material wastes are as shown in Tables 1.4–1.7, respectively.

TABLE 1.4 Requirement of Different Permissions by B1 Metal-Wastes

Code of B1 Metals and Metal-Bearing Wastes	Description
B1010	Metals and metal-alloy wastes in metallic and non-dispersible forms: • Aluminum-scraps; • Bismuth-scraps; • Chromium-scraps; • Cobalt-scraps; • Copper-scraps; • Germanium-scraps; • Iron and steel-scraps; • Magnesium-scraps; • Manganese-scraps; • Molybdenum-scraps; • Nickel-scraps; • Precious-metals like silver, gold, platinum-groups but not mercury; • Rare earth-scraps; • Scraps of indiums, hafnium, niobium, rhenium, and gallium; • Tantalum-scraps; • Thorium-scraps; • Tin-scraps; • Titanium-scraps; • Tungsten-scraps; • Vanadium-scraps; • Zinc-scraps; • Zirconium-scraps.
B1020	Clean and un-contaminated metal-scraps inclusive of alloys in bulk finished-forms such as sheets, plates, rods, beams, etc., made of the following: • Antimony-scraps; • Cadmium-scraps; • Lead-scraps exclusive of lead-acid batteries; • Tellurium-scraps.

TABLE 1.4 *(Continued)*

Code of B1 Metals and Metal-Bearing Wastes	Description
B1030	Refractory-metals that contain residues
B1031	Tungsten, molybdenum, tantalum, titanium, rhenium, and niobium metals and metal-alloy wastes in metallic dispersible-forms (metal-powders) that exclude such wastes as specified in 'list-A under entry A-1050,' for galvanic-sludges
B1040	Scrap-assemblies of un-contaminated from electrical power generation, with lubricating-oils, PCBs or PCTs to hazardous extents
B1050	Mixture of nonferrous-metals, heavy-fraction scraps, in concentrations as specified in "Schedule 2 of the Act"
B1060	Waste tellurium and selenium in metallic elemental-forms inclusive of powders
B1070	Waste of coppers as well as copper-alloys in dispersible-forms
B1080	Zinc-ashes as well as residues including zinc-alloy residues in dispersible-form unless these contain the constituents as specified in "Schedule 2 of the rules"
B1090	Waste-batteries that conform to a specification exclusive of those made with leads, cadmium or mercury
B1100	Metal-bearing wastes that arise from smelting, melting, and refining of metals as follows: • Hard-zinc spelters; • Zinc-dross: o Galvanizing-slab zinc top-dross (more than 90% of Zn); o Galvanizing-slab zinc bottom-dross (more than 90% of Zn); o Zinc die-casting dross (more than 85% of Zn); o Hot-dip galvanizers slab-zinc dross (more than 90% of Zn); o Zinc-skimming. • Aluminum-skimming exclusive of salt-slags; • Slags of copper-processing for further processing or refining that do not contain leads, arsenic or cadmium; • Wastes of refractory-linings inclusive of crucibles that originate from copper-smelting; • Slags of precious-metals processing for refining further; • Tantalum bearing tin-slags with lesser than 0.5% of tin.
B1110	Electrical and electronic assemblages:

TABLE 1.4 *(Continued)*

Code of B1 Metals and Metal-Bearing Wastes	Description
	• Electronic assemblages that consist of only metals or alloys. • Waste electrical and electronic assemblages or scraps (inclusive of printed circuit-boards) that do not contain components such as mercury-switches, accumulators, and other batteries contained on list-A, glasses from cathode ray tubes and other activated-glasses and PCB capacitors, or which are not contaminated with constituents like leads, cadmiums, mercury, polychlorinated-biphenyl) or from which these are removed to an extent of not possessing any of the characteristics contained in "Schedule 2 of the rules." • Electrical and electronic assemblages (inclusive of printed circuit-boards, electronic-components, and wires) meant for direct-reuses, and not to recycle or final disposals.
B1120	Spent-catalysts exclusive of liquids utilized as catalysts that contain any of the following: • Chromium; • Cobalt; • Coppers; • Hafnium; • Iron; • Manganese; • Molybdenum; • Nickel; • Niobium; • Rhenium; • Scandium; • Tantalums; • Titanium; • Tungsten; • Vanadium; • Yttrium; • Zinc; • Zirconium; Lanthanides (rare-earth metals): • Cerium; • Dysprosium;

TABLE 1.4 *(Continued)*

Code of B1 Metals and Metal-Bearing Wastes	Description
	• Erbium; • Europium; • Gadoliniums; • Holmium; • Lanthanum; • Lutetium; • Neodymium; • Praseodymium; • Samarium; • Terbium; • Thulium; • Ytterbium.
B1130	Cleaned-spent precious-metal bearing-catalysts.
B1140	Precious-metal bearing-residues in solid-forms containing traces of inorganic-cyanides.
B1150	Precious-metals and alloy-wastes (silver, gold, the platinum groups not including mercury) in dispersible as well as non-liquid forms having suitable packaging and labeling.
B1160	Precious metal-ashes from incineration of printed circuit boards.
B1170	Precious metal-ashes from incineration of photographic-films.
B1180	Waste photographic-films that contain silver-halides and metallic-silver.
B1190	Waste photographic-papers that contain silver-halides and metallic-silver.
B1200	Granulated-slags that are resulted from steel and iron manufactures.
B1210	Slags that are resulted from steel and iron manufactures inclusive of slags as vanadium and TiO_2 source.
B1220	Slags from production of zinc that are chemically-stabilized with a higher iron-content (>20%), and are processed in accordance with industrial-specifications (e.g., DIN-4301) mostly used for constructions.
B1230	Mill-scaling that arises from steel and iron manufactures.
B1240	Copper-oxide mill-scales.

TABLE 1.5 Requirement of Different Permissions by B2 Metal-Wastes

Code of B2 Metal-Wastes	Description
B2010	Wastes generated from mining-operations in nondispersible-forms as follows: • Natural graphite-wastes; • Slate-wastes either trimmed-roughly or merely-cut by sawing or otherwise; • Mica-wastes; • Leucites, nephelines, and nepheline-syenites wastes; • Feldspar-wastes; • Fluorspar-wastes; • Silica-wastes in solid-form exclusive of those utilized in foundry-operations.
B2020	Glass-wastes in nondispersible-forms as follows: • Cullets and other-wastes and glass-scraps exclusive of "Glasses from cathode ray tubes and other activated-glasses"
B2030	Ceramic-wastes in nondispersible-forms as follows: • Cermet-wastes and scraps (metal ceramic-composites); • Ceramic-based fibers not included or specified elsewhere.
B2040	Other-wastes that contain mainly the inorganic-constituents as follows: • Partly refined calcium-sulfates that are produced from "flue gas desulfurization (FGD);" • Waste-gypsum wall-boards or plaster-boards that arise because of building-demolitions; • Slags from production of coppers that are chemically-stabilized with a higher iron-content (>20%), and are processed in accordance with industrial-specifications (e.g., DIN-4301 and DIN-8201) mostly used for constructions and abrasive-applications; • Limestones from calcium-cyanamide productions with a pH value of lesser than 9; • Potassium, sodium, calcium-chlorides; • Carborundum (silicon-carbides); • Broken-concretes; • Lithium-tantalums and lithium-niobium that contain glass-scraps.
B2050	Fly-ashes from coal-fired power-plants not incorporated in list-A
B2060	Spent activated carbons that are resulted from the treatment of potable-water treatments and processing of the food-industries and vitamin-productions
B2070	Calcium-fluoride sludges
B2080	Waste-gypsums that arise from processing of chemical-industries not incorporated in "Schedule 2 of the rules"

TABLE 1.5 *(Continued)*

Code of B2 Metal-Wastes	Description
B2090	'Waste anode-butts' from production of steels or aluminum that are made of petroleum-cokes or bitumen, and are cleaned to normal industry-specifications (exclusive of anode-butts from chlor-alkali electrolysis and from metallurgical-industries)
B2100	'Waste-hydrates' of aluminum, waste-alumina, and residues from production of alumina exclusive of such materials utilized for gas-cleaning, flocculation or filtration-processes
B2110	'Bauxite-residues (red-mud)' with pH value to be moderated to lesser than 11.5
B2120	'Waste acidic/basic solutions' with pH value more than 2 and lesser than 11.5, which are not-corrosive or hazardous otherwise
B2130	'Bituminous-materials (asphalt-wastes)' from road maintenance and constructions that do not contain tar

TABLE 1.6 Requirement of Different Permissions by B3 Metal-Wastes

Code of B3 Metal-Wastes	Description
B3010	Solid plastic-wastes: The following plastics or mixed-plastic materials and are not mixed with other wastes are prepared to a specification: • Scrap-plastics of non-halogenated polymers and copolymers as follows: o Acrylic-polymers; o Acrylonitrile; o Alkanes C10-C13 (plasticizer); o Butadiene; o Ethylene; o Polyacetals; o Polyamides; o Polybutylene-terephthalates; o Polycarbonates; o Polyether; o Polyethylene-terephthalates; o Polymethyl-methacrylate; o Polyphenylene-sulfides; o Polypropylenes; o Polysiloxane; o Polyurethanes that do not contain CFCs; o Polyvinyl-acetates; o Polyvinyl-alcohols;

TABLE 1.6 *(Continued)*

Code of B3 Metal-Wastes	Description
	o Polyvinyl-butyral; o Styrene. • Cured waste-resins or condensation-products inclusive of the following: o Alkyd-resins; o Epoxy-resins; o Melamine formaldehyde-resins; o Phenol formaldehyde-resins; o Polyamides; o Urea formaldehyde-resins. • Fluorinated-polymer wastes as follows: o Perfluoro-ethylene/propylene (FEP); o Perfluoro alkoxyl-alkane; o Tetrafluoro-ethylene/perfluoro vinyl-ether (PFA); o Tetrafluoro-ethylene/perfluoro methylvinyl-ether (MFA); o Poly-vinyl-fluoride (PVF); o Poly-vinylidene-fluoride (PVDF).
B3130	Waste-polymer ethers and waste non-hazardous monomer-ethers unable to form peroxides
B3140	Waste-pneumatic tires

TABLE 1.7 Requirement of Different Permissions by B4 Metal-Wastes

Code of B4 Metal-Wastes	Description
B4010	Wastes comprising of primarily water-based or latex paints, inks, and hardened-varnishes that do not contain organic-solvents, heavy-metals or biocides to hazardous extents
B4020	Wastes from producing, formulating, and utilizing of resins, plasticizers, latex, glues/adhesives that are not listed in list-A, and are free of solvents as well as other-contaminants to an extent of not exhibiting the characteristics like water-based or glues-based on caseins, polyvinyl-alcohols, starches, cellulose-ethers, and dextrin
B4030	Used single-handed cameras with batteries that are not included in list-A

The list of characteristics if found present requiring the 'Central-Government Consent' before importing or exporting are as shown in Table 1.8.

TABLE 1.8 Characteristics Requiring the 'Central Government' Consent Before Importing or Exporting

Code	Characteristic
H-1	**Explosives:** These substances or wastes are solid or liquid substances or wastes (or mixture of wastes or substances) that are capable of producing gases by chemical reactions at defined pressure and temperature, and at such speeds so as to cause damages to the surroundings.
H-3	**Flammable-Liquids:** These are having the same meaning as inflammables. These are in the form of liquids, or combinations of liquids, or liquids with solids in solution or suspension, for example, varnishes, paints, lacquers, etc., but do not include wastes or substances classified on the basis of their dangerous-characteristics) which gives flammable vapors at temperatures not more than 60.5°C-at 'closed-cup test,' or not more than 65.5°C-at 'open-cup test.'
H-4.1	**Flammable-Solids:** These are the solids or waste-solids other than the explosives that are readily combustible, or may contribute to fire through frictions.
H-4.2	**Wastes or Substances Liable to Spontaneous-Combustion:** Wastes or substances which are liable to spontaneous-heating under normal-conditions encountered in transport, or get heated up on air-contact and being liable to catch-fire.
H-4.3	**Wastes or Substances Emitting Flammable Gases in Contact with Water:** Wastes or substances which are liable to be spontaneously-flammable by interacting with water, or give-off flammable-gases in dangerous-quantities.
H-5.1	**Oxidizing:** Wastes or substances are not necessarily combustible, but may yield oxygen to cause or contribute to the combustion of other materials.
H-5.2	**Organic-Peroxides:** Organic wastes or substances containing the 'bivalent –O-O-structure' are thermally-unstable substances that may undergo exothermic self-accelerating-decompositions.
H-6.1	**Acute-Poisons:** Wastes or substances which are liable to either cause death or serious-injury or harm-health, if swallowed or by skin-contacts.
H-6.2	**Infectious-Substances:** Wastes or substances that contain visible-microorganisms or their toxins which are recognized or suspected in causing diseases in human beings or animals.
H-8	**Corrosives:** Wastes or substances that cause severe damages by chemical-actions, when remain in contact with living-tissues, or in case of leakages, these cause materially-damages or even obliterate other goods or may also cause other hazards by the means of transport.
H-10	**Toxic-Gasses Liberation in Contact with Water or Air:** Wastes or substances that are liable to give-off toxic-gases in dangerous-quantities when interacted with water or air.
H-11	**Toxic (Chronic or Delayed Effects):** Wastes or substances that may involve in chronic or delayed effects including 'carcinogenicity,' when these are ingested or inhaled or if penetrated into the skins.
H-12	**Ecotoxic Effects:** Wastes or substances which are if released, then these may result in instant or delayed adverse-impacts to the environment by means of bio-accumulation or/and toxic-effects upon the biotic-systems.
H-13	This refers to the capability by any means that may cause after disposals, to yield another material of possessing any of the above-listed characteristics.

Standards as well as the technical-specifications for bio-wastes disposal and treatment are as follows:

1. **Standards for Incinerators:** All incinerators should conform to the following operating and emissions standards:
   i. **Operating-Standards:**
      a. Combustion-efficiency (CE) should be of at least 99.00%.
      b. The combustion-efficiency is to be calculated as follows:

      $$CE = \%CO_2/(\%CO_2 + \% CO) \times 100$$

      c. The primary-chamber temperature should be $800 \pm 50°C$.
      d. The secondary-chamber gas residence-time should be at least 1 second at $1050 \pm 50°C$ with minimum-oxygen in the stack-gas to be 3%.

   ii. **Emission-Standards:**
      a. Particulate-matters should have concentration mg/Nm^3 at 12% of CO_2 correction, to be 150.
      b. Nitrogen-oxides should have concentration mg/Nm^3 at 12% of CO_2 correction, to be 450.
      c. HCI should have concentration mg/Nm^3 at 12% of CO_2 correction, to be 50.
      d. Minimum stack-height should be 30 meters above ground level.
      e. Volatile organic-compounds in ashes should not exceed 0.01%.

Moreover, it may also be noted that:

- The pollution control-devices that are properly designed should be retrofitted/installed with the incinerators in order to achieve the above emission-limits, if necessary.
- Wastes should not be treated chemically with any chlorinated-disinfectants, which need to be incinerated.
- Incineration of ‘chlorinated-plastics’ should be avoided.
- Toxic-metals in incineration-ashes should be limited within the regulated-quantities as defined under the “Hazardous Waste (Management and Handling Rules, 1989).”
- Only low Sulphur-fuel like “L.D.0dLS.H.S.1 diesel” should be used as fuel in the incinerators.

2. **Standards for Waste-Autoclaving:** The autoclaves should be devoted for the purposes of disinfection and treatment of bio-medical wastes:
   i. When operating gravity-flow autoclaves, the medical-wastes should be subjected to:
      a. A temperature of not lower than 121°C and pressure of 15 pounds-per-square-inch for an autoclave residence-time of not below '60 minutes';
      b. A temperature of not lower than 135°C and a pressure of 31 pounds-per-square-inch for an autoclave residence-time of not below '45 minutes';
      c. A temperature of not lower than 149°C and a pressure of 52 pounds-per-square-inch for an autoclave residence-time of not below '30 minutes.'
   ii. While operating vacuum-autoclaves, the medical-wastes should be subjected to a minimum of one prevacuum-pulse to purge the autoclaves of all-air. The wastes should be subjected to the following:
      a. A temperature of not lower than 121°C and pressure of 15 pounds-per-square-inch for an autoclave residence-time of not below '45 minutes';
      b. A temperature of not lower than 135°C and pressure of 31 pounds-per-square-inch for residence-time of an autoclave not below '30 minutes.'
   iii. Medical-wastes should not be considered to be properly-treated unless the time-period, pressure, and temperature indicators indicate their appropriate reaches during the autoclave-process. In case of in-appropriate indications, the entire-load of medical-wastes need to be autoclaved further till the achievement of proper pressure, temperature as well as residence-time.
   iv. Each autoclave should have computer or graphic recording-devices for automatic and continuous monitoring and recording of time, date, 'load identification-number,' and 'operating-parameters' throughout the autoclave-cycle.
   v. Validation-tests: The autoclaves should kill the approved biological-indicator completely and consistently at the maximum design-capacity of each autoclave-unit. Bacillus-stea-

rothermophilus spores act as the biological-indicator for an autoclave that uses vials or spore-strips having at least '1×104-spores' per-milliliter. Autoclaves should not be allowed to have minimum operating-parameters below a residence-time of 30 minutes, at temperature or pressure below 121°C and 15 pounds-per-square inch.

vi. Routine-tests: A chemical-indicator tape/strip that changes color by reaching at a certain temperature can be used in the verification of the achievement of a specific temperature. At different locations, it may be required for using more than a single strip over the waste-packages in order to ensure that the inner-content of the package has been adequately-autoclaved.

3. **Standards for Liquid-Wastes:** The effluents that are generated from the hospitals should conform to the following limits:
   i. pH value should be of permissible-limits of '6.3–9.0.'
   ii. Suspended-solids should have permissible-limit of '100 mg/l.'
   iii. Oils and greases should have permissible-limit of '10 mg/l.'
   iv. BOD should have permissible-limit of '30 mg/l.'
   v. COD should have permissible-limit of '250 mg/l.'
   vi. Bio-assay test should cause of 90% survival of fishes after 96 hours in 100% effluents.

   It may further be noted that these limits are applicable to the hospitals having either sewers connections without terminal-sewage treatment-plants or are not connected to public-sewers. In order to release into public-sewers with terminal-facilities, the notification of general-standards under the "Environment (Protection) Act-1986," remain applicable.

4. **Microwaving Standards:**
   i. Microwave-treatment should not be used for cytotoxic, harmful, and/or radioactive wastes, contaminated animals' carcasses, body-parts, and large-metal items.
   ii. The microwave-systems should conform to the efficacy-tests or routine-tests and the supplier may require to provide a performance-guarantee before operating of the limits.
   iii. The microwaves should kill the bacteria in addition to other pathogenic-organisms completely as well as consistently, which is ensured by approved biological-indicators at

the maximum design-capacity of each microwave-units. Biological-indicators for microwave need to be "Bacillus Subtilis spores" using vials or spore-strips with at least 1 × 101 spores-per-milliliter.

5. **Standards for Deep-Burials:**
   i. A two-meters deeper trench or pit need to be digged, which should be half-filled with wastes followed by covering with limes within 50 cm of the surface prior to filling of the rest pit with soils.
   ii. Animals should not have any access to burial-sites. For this, covers of galvanized-iron/wire meshes may be utilized.
   iii. A layer of 10 cm of soils need to be for covering the wastes, for each event of wastes addition to the pit.
   iv. Under close-supervisions, the burials need to be performed.
   v. Deeper burial-sites should be relatively impervious and thus, no shallow-wells should be closer to the sites.
   vi. There must be maintained distance between pits and habitation for ensuring of non-contamination of any surface or ground-water. The sites should not be prone to erosions or flooding.
   vii. Prescribed authority must have the authorization of the location of the deep burial-sites.
   viii. The institutions must maintain records of all pits for deep-burials.

Segregating wastes at the source and safe-storage is the key to the whole waste management process generated from hospitals. Therefore, segregating various wastes into different categories in accordance with the treatment or disposal options need to be carried out at the generation-points in color-coded plastic containers/bags as per schedule-II of the gazette-notification. The syringes and needles need to be mutilated and disinfected before segregations. The containers types' and their color-codes as set by Government of India notifications are as illustrated in Table 1.9.

The state-wise status of MSW processing-facilities during 2011 in India (Planning Commission, 2014) is as illustrated in Table 1.10.

The landfill-sites associated with different cities in India (Parvathamma, 2014), is as shown in Table 1.11.

Table 1.12 illustrates the state-wise installed and operational capacities of compost-plants in India (Department of Fertilizers, 2017).

TABLE 1.9 Waste-Category and Color-Code for Waste Disposal Systems

Waste-Category	Type of Container	Color-Code
1, 2, 3 and 6	Plastic-hags	Yellow
4 and 7	Puncture-proof containers or plastic bags	White/blue translucent
3, 6 and 7	Disinfected containers or plastic-bag	Red
5, 9 and 10 (solids)	Plastic-bags	Black

Notes:
- The color-coding of waste-categories with options in multiple treatment as defined in schedule I need to be selected depending on chosen treatment options.
- Chlorinated plastics need to be avoided to make the waste-collecting bags.
- Any bags or containers are not required for "categories 8 and 10 (liquid)."
- No requirement of putting in bags/containers for "category 3," if disinfected, locally.

TABLE 1.10 State-Wise Status of MSW Processing-Facilities During 2011 in India

State	Compos-ting	Vermi-composting	Bio-Metha-nation	Pelletization	Waste-to-Energy
Assam	1	Nil	Nil	Nil	Nil
Andhra Pradesh	24	Nil	Nil	11	2
Chhattisgarh	6	Nil	Nil	Nil	Nil
Chandigarh	Nil	Nil	Nil	1	Nil
Delhi	3	Nil	Nil	Nil	3
Gujarat	3	93	Nil	6	Nil
Himachal Pradesh	10	Nil	Nil	Nil	Nil
Jharkhand	4	Nil	Nil	Nil	Nil
Kerala	21	7	10	1	1
Maharashtra	6	2	5	5	2
Madhya Pradesh	7	Nil	Nil	2	Nil
Nagaland	1	1	Nil	Nil	Nil
Odisha	1	Nil	Nil	Nil	Nil
Punjab	1	3	Nil	Nil	Nil
Sikkim	1	Nil	Nil	Nil	Nil
Tamil Nadu	162	24	Nil	3	Nil
West Bengal	13	7	Nil	Nil	Nil

TABLE 1.11 Landfill-Sites Associated with Different Cities in India

City	Number of Landfills (Area of Landfills in Hectare)
Ahmadabad	1 (84)
Bangalore	2 (40.7)
Chandigarh	1 (18)
Coimbatore	2 (292)
Chennai	2 (465.5)
Delhi	3 (66.4)
Guwahati	1 (13.2)
Hyderabad	1 (121.5)
Indore	1 (59.5)
Jabalpur	1 (60.7)
Jaipur	3 (31.4)
Kolkata	1 (24.7)
Kanpur	1 (27)
Ludhiana	1 (40.4)
Mumbai	3 (140)
Madurai	1 (48.6)
Meerut	2 (14.2)
Nasik	1 (34.4)
Raipur	1 (14.6)
Ranchi	1 (15)
Surat	1 (200)
Srinagar	1 (30.4)
Thiruvananthapuram	1 (12.5)
Vishakhapatnam	1 (40.5)

TABLE 1.12 State-Wise Installed and Operation-Capacities of Compost-Plants in India

State	Number of Plants	Installed-Capacity in Tons Per Year (Operation-Capacity in %)
Andaman and Nicobar Islands	1	90 (–)
Andhra Pradesh	2	2,400 (20)
Assam	1	15,000 (15)
Chhattisgarh	1	1,200 (20)
Daman and Diu	1	4,050 (–)
Delhi	4	1,68,000 (16.1)
Goa	1	1,200 (20)

TABLE 1.12 *(Continued)*

State	Number of Plants	Installed-Capacity in Tons Per Year (Operation-Capacity in %)
Gujarat	15	1,74,300 (19.5)
Haryana	4	18,600 (15.3)
Karnataka	18	4,73,400 (10.1)
Kerala	3	1,56,000 (20)
Madhya Pradesh	1	36,000 (15)
Maharashtra	13	4,88,400 (12.5)
Punjab	2	19,200 (15)
Rajasthan	1	1,80,000 (15)
Tamil Nadu	9	67,680 (15.8)
Telangana	5	1,92,000 (15)
Tripura	1	75,000 (6)
Uttar Pradesh	7	1,24,560 (15.2)
West Bengal	5	1,70,400 (15)
Total	95	23,67,480 (14)

Various allocations of lands to develop landfills in India (Central Pollution Control Board, 2011), is as shown in Table 1.13.

TABLE 1.13 Allocations of Lands to Develop Landfills in India

City	Number of Landfill-Sites (Areas in Acres)
Surat	1 (494.2)
Hyderabad	1 (300.2)
Ahmedabad	1 (207.6)
Delhi	3 (164.1)
Jabalpur	1 (150.7)
Indore	1 (147)
Madurai	1 (120.1)
Bengaluru	2 (100.6)
Vishakhapatnam	1 (100.1)
Ludhiana	1 (99.8)
Nasik	1 (85)
Jaipur	3 (77.6)
Srinagar	1 (75.1)
Kanpur	1 (61)

TABLE 1.13 *(Continued)*

City	Number of Landfill-Sites (Areas in Acres)
Kolkata	1 (61)
Chandigarh	1 (44.5)
Ranchi	1 (37.1)
Raipur	1 (36.1)
Meerut	2 (35.1)
Guwahati	1 (32.6)
Thiruvananthapuram	1 (30)
Vadodara	1 (20)
Dehradun	1 (11.1)
Jamshedpur	2 (10.1)
Faridabad	3 (5.9)
Asansol	1 (4.9)
Varanasi	1 (4.9)
Agra	1 (3.7)
Lucknow	1 (3.5)
Rajkot	2 (3)
Shimla	1 (1.5)
Mumbai	3 (345.9)
Chennai	2 (1150.3)
Coimbatore	2 (721.5)

The organic residues along with the related effluents in a huge amount are produced every year by the food-processing industries like meat, chips, juice, confectionery, and fruit-industries. The organic residues generated through these industries can be utilized for different energy-sources. With the continuous increase in population, the food requirements and their utilization also increased. As a result, different food and beverage producing industries have increased remarkably in different regions in most of the countries in order to fulfill the need of foods. Different compositions of fruit-industrial wastes are illustrated in Table 1.14, which comprise of different compositions of hemi-cellulose, cellulose, lignin, ash, moisture, nitrogen, carbon, etc. The production of vegetables and fruits in India that count for an approximate amount of 20% to be going as wastes every-year (Rudra et al., 2015). As a large amount of apples, cottons, soy-beans, and wheats are produced in India. So, it also increased the wastes' percentage produced from them.

TABLE 1.14 Composition of Fruit-Industrial Wastes

Fruit-Industrial Waste	Chemical-Composition in % w/w							References
	Hemi-Cellulose	Cellulose	Lignin	Ash	Moisture	Nitrogen	Carbon	
Orange peel	10.5	9.21	0.84	3.5	11.86	–	–	Rivas et al. (2008)
Potato peel waste	–	2.2	–	7.7	9.89	–	1.6	Weshahy and Rao (2012)
Pineapple peel	–	18.11	1.37	–	91	0.99	40.8	Paepatung et al. (2009)
Coffee skin	16.68 (g/100 g)	23.77 (g/100 g)	28.58 (g/100 g)	5.36 (g/100 g)	–	–	C/N 14.41	Lina et al. (2014)

The nutrient-potentials of various crop-residues in India (Tandon, 1997), are illustrated in Table 1.15.

TABLE 1.15 The Nutrient-Potentials of Various Crop-Residues in India

Crops	Percentage of Contents				Tons of Nutrients Per Tons of Residues
	N	P_2O_5	K_2O	Total	
Maize	0.52	0.18	1.35	2.05	0.0205
Wheat	0.48	0.16	1.18	1.82	0.0182
Rice	0.61	0.18	1.38	2.17	0.0217
Potato-tubers	0.52	0.20	1.06	1.79	0.0179
Pearl-millets	0.45	0.16	1.14	1.75	0.0175
Pulses	1.29	0.36	1.64	3.29	0.0329
Sugarcanes	0.40	0.18	1.28	1.86	0.0186
Sorghum	0.52	0.23	1.34	2.09	0.0209
Groundnuts	1.60	0.23	1.37	3.20	0.0320
Finger-millet	1.00	0.20	1.00	2.20	0.0220
Barley	0.52	0.18	1.30	2.00	0.0200
Oil-seeds	0.80	0.21	0.93	1.94	0.0194

The toxicity of the biological systems by heavy metals are higher and the risk-level is a function of the contaminant quantity in addition to the exposed time of the organism (Goyer, 1996). Table 1.16 shows the heavy metal ions, the major pollution sources and the toxicological-characteristics caused by exposure to As, Cd, Cr, Cu, Zn, and Pb. The Pathogenic-organisms causing infections depend on the organisms' resistance to sewage treatments, environmental conditions, infection doses, pathogenicity, degree of host-immunity, susceptibility, and degree of human-exposures to the foci of transmissions (CETESB, 1993). The major presence of pathogens in the sludge are parasites, bacteria, and viruses (CONAMA, 2006), which is as summarized in Table 1.17. As the quantity of these microorganisms is variable and depends on the time as well as season of the year. Thus, in order to utilize the sludge in agriculture, it is necessary for characterization and quantifying the present chemical-contaminants and pathogenic microorganisms (CETESB, 1991; Rodrigues, 2006). Consequently, the systematic treatment of the sewage-sludges or urban-wastes before utilization in agricultural-soils diminishes the risks to human and animal health by infections as it reduces the survival chances of these pathogenic-organisms.

TABLE 1.16 Heavy Metal-Ions, the Major Pollution Sources and the Toxicological-Characteristics

Ions (Symbol)	Major Pollution Sources	Toxicological-Characteristics
Arsenic (As)	• Insecticides, herbicides, fungicides, glassware, and mining. • Paint and dyes industries.	The inorganic compounds are more toxic. As^{3+} is more toxic than As^{5+} at very high-doses. Affected organs and tissues are: skins, respiratory-systems, cardiovascular-systems, reproductive-systems, gastrointestinal-systems, and nervous-systems (IPCS-International Program on Chemical Safety, 2001)
Cadmium (Cd)	Industrial-effluents, pigments production, electroplating, electronic-equipment, lubricants, photographic-accessories, insecticides, and fossil-fuels.	In rat studies, the respiratory tract of the animals was found to be exposed to continuous aerosol with a lower concentration of $CdCl_2$, and a higher incidence of lung cancer was observed. Higher levels of Cd inhalations cause lethal-pulmonary-edema (IPCS-International Program on Chemical Safety, 1992)
Chromium (Cr)	Industrial-effluents, aluminum, and steel production, pigments, paints, explosives, papers, and photography.	The toxic-effects of Cr^{3+} occur only through parenteral-administration. The development of cancer occurs when humans and other animals are exposed to Cr. The gastrointestinal tract is affected by Cr^{6+} in the diet, which further affects the kidneys, and the hematological-systems, and causes several genetic-damages. The $CrCl_3$ was found to be accumulated in the cell nucleus, in some studies (IPCS-International Program on Chemical Safety, 1988; USEPA-United States Environmental Protection Agency, 1978)
Copper (Cu)	Pipe-corrosions, domestic-sewages, fungicides, algicides, pesticides, foundries, mining, and metal-refinements.	Acute effects related few cases of Cu have been reported, that include gastric-burning, nausea, vomiting, diarrheas, hemolytic-anemia, and lesions of the gastrointestinal tract. Chronic-effects are reported rarely, except for "Wilson's disease," which is responsible for the accumulation of coppers in the liver, kidney, and brain (Quinráglia, 2001)
Zinc (Zn)	Waste-incineration, mining, combustion of woods, electroplating, domestic-sewages, and steel and iron productions.	As the accumulations of Zn do not cause any profound deficiencies, so it is considered to have lower-toxicity. Excessive intake of Zn may cause gastrointestinal-disorders and diarrhea (Eleutério, 1977)

TABLE 1.16 *(Continued)*

Ions (Symbol)	Major Pollution Sources	Toxicological-Characteristics
Lead (Pb)	Industrial-effluents, paints, tobacco, pipes, metallurgical, and electrodeposition industries.	After the absorption of Pb by human-body, it can be found in the bloods, soft, and mineralized-tissues (ATSDR, 1999). According to ATSDR (1992), for neurological, metabolic, and behavioral reasons, the children are more vulnerable to the Pb effects than the adults, which include weakest intelligence-quotient, effects on the nervous-systems, reductions in sensory-functions, involuntary functions of kidney and nervous systems, and premature-births (IPCS, 1995; Vega-Dienstmaier et al., 2006; Bellinger, 1995)

TABLE 1.17 Organisms Present in Sewage-Sludges and Risky to Human-Health

	Organisms	Disease and Symptoms
Bacteria	*Salmonella* sp.	*Salmonellosis* (typhoid-fever)
	Shigella sp.	Bacillary-dysentery
	Campylobacter jejuni	Gastroenteritis
	Vibrio cholerae	Cholera
	Escherichia coli patogênica	Gastroenteritis
Enterovirus	Polioviruses	Poliomyelitis
	Coxsackievirus	Meningitis, hepatitis, pneumonia, and fever
	Reoviruses	Respiratory-infections, gastroenteritis
	Caliciviruses	Gastroenteritis
	Astroviruses	Gastroenteritis
Enteric viruses	Rotaviruses	Acute-gastroenteritis
	Hepatitis-A	Infectious-hepatitis
	Norwalk	Severe diarrhea with gastroenteritis
Helminths	*Necator americanus*	Ancylostomiasis
	Taenia saginata	Anorexia, nervousness, abdominal-pain, digestive-disorders
	Toxocara canis	Abdominal-discomfort, muscle-pains, neurological-symptoms
	Trichuris trichiura	Abdominal-pain, anemia, diarrhea, weight-losses
	Ascaris lumbricoides	Digestive-problems, nutritional-disorders, and abdominal-pains
Protozoa	*Balantidium coli*	Diarrhea
	Entamoeba histolytica	Gastroenteritis
	Toxoplasma gondii	Toxoplasmosis
	Cryptosporidium	Gastroenteritis
	Giardia lamblia	Giardia (Diarrhea, abdominal pain, and weight-losses)

The options of waste management with key advantages and disadvantages as suggested by Rushton (2003) are illustrated as follows:

1. **Recycling:**
   - Advantages include:
     - o Resources conservation;
     - o Industrial raw materials supply;
     - o Reduction of disposed wastes to landfills and incineration.

- **Disadvantages include:**
  - o A variety of processes;
  - o Emissions from recycling-processes;
  - o More energy utilization for processes as compared to original manufacturing;
  - o Lower demand for products in the present scenario;
  - o Requirement of individuals co-operation.

2. **Composting:**
   - Advantages include:
     - o Decreased disposals of wastes to landfills and incineration;
     - o Recovery of useful organic-matters for use as soil-amendments;
     - o Employment-opportunities.
   - Disadvantages include:
     - o Noises, odors, vermin-nuisance;
     - o Bio-aerosols-organic dusts comprising bacteria or fungal-spores;
     - o Emission of volatile organic compounds;
     - o Potential-pathway from use-on-land for contaminants to enter food-chains.

3. **Sewage Treatment:**
   - Advantages include:
     - o Safer-disposal of human-wastes;
     - o Protection of potable water-supply sources.
   - Disadvantages include:
     - o Discharges may have organic-compounds, endocrine-disrupting compounds, heavy-metals, pathogenic-microorganisms;
     - o Odor-nuisance.

4. **Incineration:**
   - Advantages include:
     - o Reduction of weights and volume of wastes, and about 30% remain as ash that can be used for materials-recovery;
     - o Reduction of potential-infectivity of clinical-wastes;
     - o Production of energy for electricity generation.
   - Disadvantages include:
     - o Production of hazardous solid-wastes;

- o Discharges contain contaminated waste-water;
- o Emission of toxic-pollutants, heavy-metals, and combustion-products.

5. **Landfill:**
    - Advantages include:
        - o Cheaper-disposal method;
        - o Wastes are used to back-fill-quarries before reclamation;
        - o Landfill-gases contribute to renewable-energy supply.
    - Disadvantages include:
        - o Water-pollution from leachate and run-off;
        - o Air-pollution from anaerobic-decomposition of organic-matters to produce carbon dioxide, methane, sulfur, nitrogen, and volatile organic compounds;
        - o Emits known or suspected teratogens or carcinogens (e.g., nickel, arsenic, chromium, benzene, dioxins, vinyl-chloride, polycyclic-aromatic hydrocarbons);
        - o Animal-vectors (seagulls, rats, flies) for a few diseases;
        - o Dust, odor, road traffic-problems.

Moreover, the list of waste prevention measures that could be incorporated in the local wastes' management are as follows:

1. Planning wastes-prevention (Cox et al., 2010; Phillips et al., 2011; Thyberg and Tonjes, 2015; Zorpas et al., 2015) that includes:
    - Leaderships;
    - Improvement in organizational-capacities, augmented knowledge;
    - Specific reduction-targets' setting for the residual-wastes;
    - Base-plan on reliable-data and area-specific features for waste compositional-analysis;
    - Data-collection, performance-evaluation;
    - Review continuation and improvement;
    - Stakeholders' engagement;
    - Communications in addition to outreach;
    - Zero-waste places;
    - Linkage to other initiatives.
2. Campaigns (Cox et al., 2010; Phillips et al., 2011; Williams et al., 2015; Zorpas et al., 2015) that includes:
    - Differentiate between preventions and recycling (the drivers for sorting-behavior and prevention-behavior are different);

- Addresses the variety of practices causing wastes (cooking, shopping, food storage, etc.);
- Focuses on single specific waste-stream, or/and specific-behaviors;
- Stimulation of general environmental-attitudes, enhancement of environmental-values;
- Articulation of consumer-conflicts;
- Skill-building, tangible-tools to reduce wastes;
- Cookbook-for-leftovers;
- Training-sessions in schools;
- Meetings as well as conferences with stakeholders;
- Cloth-nappy-schemes (Bulkeley and Gregson, 2009).

3. Food-wastes campaigns (Evans, 2011, 2012; Graham-Rowe et al., 2014) that include:
   - Targeting on "waste-concerns;"
   - Highlighting the food-waste prevention-benefits, such as financial-rewards;
   - Emphasizing on reduction as the 'right' way to do (social-responsibility, ethics);
   - Involvement of consumers in difficulty by addressing complex-issues;
   - Empowerment of peoples: Skill-building.
4. Others that include (Evans, 2011):
   - Interventions in the food provisioning ways, e.g., changing in package types, sizes, and materials.

However, Giusti (2009) has discussed that "epidemiological study" is mainly concerned by defining the strength of the association between "exposures to potentially toxic-substances and specific health-effects." This is generally achieved by calculation of the ratio of the disease incidence in the exposed population over the same disease incidence in the non-exposed population, known as the "relative-risk (RR)," in some studies, also called "odd-risk (OR)." An RR > 1 indicated an increased risk of developing a specific health-outcome. For example, if the RR= 5, then the risk is five-times higher or an increase of 400%. In order to find the actual causes of a health-effect by an agent, many other issues need to be considered like the statistical significance of the association. The usual confidence-level (confidence-interval) reported is 95%, which become occasionally to 99%. The model used by the "World Cancer Research Fund" and the "American Institute for Cancer Research" (WCRF and AICR, 1997) is shown in Table 1.18, for defining the strength of evidence of an association between exposures and illnesses. Table 1.19 illustrates the examples of "RR or OD" values for the association between "environmental-factors and health," and between "exposures to facilities in wastes management and health."

TABLE 1.18 Simplified-Version of the Relative-Risk (RR) and Odd-Ratio (OR) Model

RR or OR	Statistical Significance (Strength of Evidence)
0.87–1.5	No (no-association)
1.5–2.0	No (no-association)
1.5–2.0	Yes (moderate)
>2	No (moderate)

The classifications of waste-stream by its' sources (Tchobanoglous and Kreith, 2002), is summarized in Table 1.20.

The first step to create a waste management program is to define and establish clear goals. By knowing the waste management plan strategies to achieve the goal before its design can make the process simpler. It becomes necessary to identify the goals in line with the interests as well as the core principles of any organization (USEPA, 1995).

The key objectives or goals, targets or indicators, and strategies that have been outlined in various waste management frameworks in institutional, commercial, and industrial (IC&I) sectors are summarized below:

1. **Objectives or Goals:** Minimization of waste-generations (Metro Vancouver, 2010; Environmental Defense-McDonald's Waste Reduction Task Force, 1991; University of the Sunshine Coast, 2010).
   - ➢ The targets or indicators are as follows:
     - Reducing per capita waste generation quantity of (Metro Vancouver, 2010).
     - Eliminating un-wanted materials (Environmental Defense-McDonald's Waste Reduction Task Force, 1991).
     - Systematization of solid wastes reduction and management-practices into standard operating-procedures and

TABLE 1.19 Examples of "RR or OD" Values for the Association Between "Environmental-Factors and Health," and Between "Exposures to Facilities in Wastes Management and Health"

Causes		Effects	RR, OR	CI	References
Environmental factors	Cigarettes smoking	• Liver-cancer	• 104	• 51–212	Tomatis (1990)
	• 1–14 cigarettes	• Lung-cancer	• 7.8	• NA	
	• 15–24 cigarettes	• Lung-cancer	• 12.7	• NA	
	• ≥25 cigarettes	• Lung-cancer	• 25.1	• NA	
	Hepatitis-B-virus				
Landfilling	Nant-y-Gwynedd, UK	Congenital-anomalies	• 1.9	• 1.3–2.9	Fielder et al. (2000)
		• Before opening of site, 1983–1987.	• 3.6	• 2.3–5.7	
		• First 2 years of operations, 1988–1989.	• 1.9	• 1.2–3.0	
		• From 1990 to 1996			
	9565-UK landfill sites	• Congenital-anomalies	• 1.01	• 1.01–1.02	Elliott et al. (2001a, b)
		• Low birth-weight	• 1.04	• 1.03–1.05	
		• Cardiovascular-defects	• 0.96	• 0.93–0.99	
Incineration	Besancon-incinerator, France	• Non-Hodgkin's lymphomas	• 1.27	• 1.1–1.4	Viel et al. (2000)
		• Soft tissue sarcoma	• 1.44	• 1.1–1.9	
	13 Incinerators, France	Non-Hodgkin's lymphomas	1.12	1.002–1.251	Viel et al. (2008)
	Incinerator and other industrial sources of dioxins, Italy	Sarcoma: population with the highest exposure	3.30	1.24–8.76	Zambon et al. (2007)

TABLE 1.19 *(Continued)*

Causes		Effects	RR, OR	CI	References
Sewage contaminated water	Sydney beaches, Australia	Gastroenteritis, from exposure to: • 0–39 fecal *streptococci* per 100 mL • 40–59 fecal *streptococci* per 100 mL • 60–79 fecal *streptococci* per 100 mL • 80+ fecal *streptococci* per 100 mL	• 1.0 • 1.91 • 2.90 • 3.17	• 1.60–2.28 • 1.43–5.88 • 1.12–8.97	Kay et al. (1994)
	Sydney beaches, Australia	Any symptom*, from exposure to: • 10–300 cfu/100 mL • 300–1000 cfu/100 mL • 1000–3000 cfu/100 mL	• 2.9 • 3.8 • 5.2	• 1.7–5.1 • 2.1–7.1 • 1.7–16.0	Corbett et al. (1993)
Exposure to ionizing radiations	Nuclear industry workers	• All cancers excluding leukemia • Leukemia, excluding chronic • Lymphocytic leukemia	0.97** 1.10*** 1.93** 1.19***	• 0.14–1.97 • <0–8.47	Cardis et al. (2005)

TABLE 1.16 *(Continued)*

NA = Not available.

*Fever, cough, ear or eye symptoms, gastrointestinal symptoms; cfu = colony-forming-units.

**For a radiation-dose of 1 Sv.

***For a radiation-dose of 100 mSv.

TABLE 1.20 Wastes-Streams Classification by Sources

Sources	Facilities, Activities or Locations of Wastes Generation	Solid Wastes' Types
Residential	Single-family and multi-family dwellings; apartments of low, medium, and high-density.	Food-wastes, papers, cardboards, textiles, plastics, yard-wastes, woods, ashes, street-leaves, special-wastes (e.g., bulky-items, consumer-electronics, white-goods, universal-wastes) and household hazardous-wastes.
Commercial	Markets, hotels, stores, restaurants, office buildings, motels, print-shops, service-stations, auto repair shops.	Papers, cardboard, plastics, woods, food-wastes, glasses, metal-wastes, ashes, special-wastes, hazardous-wastes.
Institutional	Universities, schools, hospitals, prisons, governmental-centers.	Same as commercial in addition to biomedicals.
Agricultural	Field as well as row crops, vineyards, orchards, dairies, farms, feedlots.	Spoiled-foods, agricultural-wastes, hazardous-wastes.
Industrial	Refineries, construction, fabrication, chemical-plants, light, and heavy manufacturing, power-plants, demolitions.	Same as commercial in addition to industrial-process wastes, scrap-materials.
Industrial (non-process wastes)	Refineries, construction, fabrication, chemical-plants, light, and heavy manufacturing, power-plants, demolitions.	Same as commercial.
Municipal solid wastes	All of the preceding.	All of the preceding.
Constructions and Demolitions	New construction-sites, road-repairs, renovation-sites, buildings razing, broken-pavement.	Woods, steels, concrete, asphalt-paving, asphalt-roofing, gypsum-boards, rocks, and soils.

product or package specifications (Environmental Defense-McDonald's Waste Reduction Task Force, 1991).
- Assessment of waste generation potentials of latest-developments (University of the Sunshine Coast, 2010).

➢ The key strategies to achieve the objectives or goals are as follows:
- Supporting in transfer of additional responsibilities of waste management to consumers and producers (Metro Vancouver, 2010).
- Reducing or eliminating the entering-materials to the solid waste system that protect the opportunities to achieve recycling, reuse, or energy-recovery (Metro Vancouver, 2010; University of Victoria, 2004).
- Providing information as well as education on waste-reducing options (Metro Vancouver, 2010).
- Evaluation of shipping and packaging procedures in order to identify the items that could be reduced or eliminated (Environmental Defense-McDonald's Waste Reduction Task Force, 1991).
- Documentation of the campus waste stream details and regular-review to assess the trends (University of the Sunshine Coast, 2010).
- Outlining the roles and responsibilities of all the involved stakeholders with waste management (Halifax Regional School Board, 2009).

2. **Objectives or Goals:** Maximization of recycling, reuse, and material-recovery (Metro Vancouver, 2010; Environmental Defense-McDonald's Waste Reduction Task Force, 1991; Nova Scotia Environment, 2009; University of Victoria, 2004).

➢ The targets or indicators are as follows:
- Increasing the waste diversion-rates (Metro Vancouver, 2010; University of Victoria, 2004).
- Using alternate-materials to reduce production-impacts (Environmental Defense-McDonald's Waste Reduction Task Force, 1991).
- Substituting reusable-items for disposable-items in storage, handling, shipping, and operations (Environmental Defense-McDonald's Waste Reduction Task Force, 1991).

- ➢ The key strategies to achieve the objectives or goals are as follows:
  - Increasing the opportunities for recycling and reuse (Metro Vancouver, 2010; Environmental Defense-McDonald's Waste Reduction Task Force, 1991).
  - Increasing the effectiveness of existing recycling-programs (Metro Vancouver, 2010).
  - Targeting specific-materials for recycling, reuse, and material-recovery (Metro Vancouver, 2010; Environmental Defense-McDonald's Waste Reduction Task Force, 1991; University of Victoria, 2004).
  - Targeting specific-waste streams (such as C&D waste) for improving the diversion-rates (Metro Vancouver, 2010).
  - Targeting specific-sectors for improving the diversion-rates (Metro Vancouver, 2010).
  - Utilizing non-recyclable materials as fuels in order to provide electricity and district-heating from waste-to-energy facilities (Metro Vancouver, 2010).
  - Developing re-usable containers for shipping purposes (Environmental Defense-McDonald's Waste Reduction Task Force, 1991).
  - Outlining the roles and responsibilities of all involved stakeholders with waste management activities (Halifax Regional School Board, 2009).

3. **Objectives or Goals:** Development of community-based waste management practices (University of the Sunshine Coast, 2010; University of Victoria, 2004).
   - ➢ The targets or indicators are as follows:
     - Developing waste management plans by consulting with the participant-groups (University of the Sunshine Coast, 2010).
     - Including communication-links, so that peoples can be informed about any change in activities having impacts on waste management (University of the Sunshine Coast, 2010).
     - Formation of a working-group for coordinating the development of specific group or area plans (University of the Sunshine Coast, 2010).

- Working with regional organizations for minimizing facilities and resources duplication (University of the Sunshine Coast, 2010).

➢ The key strategies to achieve the objectives or goals are as follows:
  - Creation of tools and materials to target community groups as well as members (University of the Sunshine Coast, 2010; University of Victoria, 2004).
  - Holding activity-sessions by detailing the importance of waste management and the roles of peoples (University of the Sunshine Coast, 2010).
  - Summary inclusion of staff-expectations in their employment-orientation (University of the Sunshine Coast, 2010).
  - Developing communication-links between among the involved groups in waste management activities. In order to avoid both overlaps and gaps, it is essential to know each other's actions in specific waste activities, such as collection, purchasing, storage, and disposals (University of the Sunshine Coast, 2010).
  - Identifying the options for cooperative product-purchasing that includes prices and discounts for bulk-purchases (University of the Sunshine Coast, 2010).
  - Inviting comments from local organizations as well as businesses (University of the Sunshine Coast, 2010).

4. **Objectives or Goals:** Adjustment in procurement-policies for better commitment to the principles of waste management (University of the Sunshine Coast, 2010).

➢ The targets or indicators are as follows:
  - Using commitment for waste management as a lobby-point for pursuing fund for capital-works (University of the Sunshine Coast, 2010).
  - Supporting the "front-end" reduction of the waste stream policy (University of the Sunshine Coast, 2010).
  - Developing regional-alliances for maximizing the purchase-power and encouraging waste avoidance-specifications for products (University of the Sunshine Coast, 2010).

- ➢ The key strategies to achieve the objectives or goals are as follows:
  - Developing purchase guidelines consistent with the waste management strategies (University of the Sunshine Coast, 2010).
  - Designing specifications for tender in such a way that the waste management issues can be properly addressed (University of the Sunshine Coast, 2010).
  - Identifying regional-bodies having similar purchase-requirements (University of the Sunshine Coast, 2010).

5. **Objectives or Goals:** Development of educational-programs (University of the Sunshine Coast, 2010; Nova Scotia Environment, 2009; University of Victoria, 2004).
   - ➢ The targets or indicators are as follows:
     - Involving the community through increasing awareness, meetings, specific information-needs as well as fostering community sense of commitments (University of the Sunshine Coast, 2010; University of Victoria, 2004).
     - Fostering competency among waste management staffs in opportunities identification for avoiding and minimizing the wastes being disposed-off currently (University of the Sunshine Coast, 2010; Halifax Regional School Board, 2009).
     - Ensuring of the operational-staffs regarding the training in order to comply with relevant legislation or guidelines, and supporting to report failures or negative-events of the system (University of the Sunshine Coast, 2010; Halifax Regional School Board, 2009).
   - ➢ The key strategies to achieve the objectives or goals are as follows:
     - Conducting studies on waste characterization to establish goals of waste-reduction (Environmental Defense-McDonald's Waste Reduction Task Force, 1991).
     - Tracking diversion-progress and making information available (Environmental Defense-McDonald's Waste Reduction Task Force, 1991).

    - Developing marketing-programs to attract participation of regional-organizations (University of the Sunshine Coast, 2010).

6. **Objectives or Goals:** Ensuring of safety and effectiveness of waste management (University of the Sunshine Coast, 2010; University of Victoria, 2004).
   - ➢ The targets or indicators are as follows:
     - Ensuring the compliances with regulations (University of the Sunshine Coast, 2010; Halifax Regional School Board, 2009).
     - Assigning responsibilities for the regular-review of the available-technologies for waste storages and disposals (University of the Sunshine Coast, 2010).
   - ➢ The key strategies to achieve the objectives or goals are as follows:
     - Documentation of the segregations, containments, storages, collections, and disposals mechanism for each waste-category with more attention to harmful-categories of wastes (University of the Sunshine Coast, 2010; University of Victoria, 2004).
     - Developing accident response-strategies for harmful-categories of wastes and providing training for the responsible peoples (University of the Sunshine Coast, 2010).
     - Providing staff-trainings (University of the Sunshine Coast, 2010; Halifax Regional School Board, 2009).
7. **Objectives or Goals:** To become a regional leader in waste management (University of the Sunshine Coast, 2010).
   - ➢ The targets or indicators are as follows:
     - Supporting initiatives at regional-levels waste management (University of the Sunshine Coast, 2010).
   - ➢ The key strategies to achieve the objectives or goals are as follows:
     - Documentation of a waste management strategy of "wish-list" including options, costs, benefits, and the parameters required to be met before the active consideration of each option (University of the Sunshine Coast, 2010).
     - Advertising the waste management initiatives, which should not be overstated, and should be included with the

limitations' discussion (University of the Sunshine Coast, 2010).
- Inviting the comments from regional businesses as well as organizations (University of the Sunshine Coast, 2010).

The differences in the two approaches of waste management such as centralized and decentralized systems of waste management is illustrated in Table 1.21. The centralized systems provide a "one-size-fits-all" solution with larger facilities for handling thousand tons-per-day, in a controlled market by few companies contracted directly by the national government. While, the decentralized approach promotes smaller-areas of services by handling lesser wastes and empowering the local authorities and communities.

TABLE 1.21 SWOT Analysis of Centralized and De-Centralized Systems of Waste Management

Centralized System	
Strengths:	**Weaknesses:**
• Longer collection distance and controlled-landfilling;	• Lower rate of diversion and treatment-quality;
• Scale economy;	• Higher inertia and ventures;
• Equality of basic-service for all.	• Established corruptions.
Opportunities:	**Threats:**
• Potential of optimization;	• Crisis repetition;
• Unique-framework and monitoring.	• Lack of change in business.
De-Centralized System	
Strengths:	**Weaknesses:**
• Source level sorting;	• Handling as well as disposal of residuals and toxic-wastes;
• Quality-treatment;	• Scale economy;
• Driving-change with user-inclusivity.	• Multiple control-facilities.
Opportunities:	**Threats:**
• Local-jobs;	• Dispersion of environmental-pollution;
• Community-empowerment;	• Corruption diffusion potential;
• Flexibilities in treatment-solutions.	• Service-inequalities;
	• Deficiency of training and expertise.

In some countries, "Life Cycle Assessment (LCA)" has been found to be applied popularly in order to manage EW (Table 1.22), which aims to measure the environmental-burdens with regard to some product, process or service by the identification of consumed-materials in addition to generated-emissions to the environment. Further, the alternative ways for improvement in the environment can be determined (Cherubini et al., 2009). Moreover, during the design phase of new electronic-products, this tool can be used to design environmental-friendly products. It can also be used for minimizing the amount of generated wastes at their "EoL." Many studies have investigated in Europe, regarding the use of LCA for the assessment of the environmental influences of EoL-treatment of EW (Kiddee et al., 2013). It has been demonstrated that the distance-traveled for the collection of EW and the treatment facilities location to be important in order to design recycling-networks and the environmental-efficiency.

TABLE 1.22 LCA Applications by Various Nations for E-Wastes

Country	Equipment: (Applications)	References
United Kingdom	Printers: (Examination of environmental-influences)	Mayers et al. (2005)
Korea	TV-sets, washing-machines, refrigerators, and air-conditioners: (Environmental and financial aspects)	Kim et al. (2004)
European countries	Washing-machines, refrigerators, TV-sets, and personal-computers: (Evaluation of environmental perspectives)	Barba-Gutiérrez et al. (2008)
Taiwan	Computers: (Evaluation of economic and environmental aspects)	Lu et al. (2006)
China	Desktops, personal-computers: (Investigation of environmental performances)	Duan et al. (2009)
Thailand	Fluorescent-lamps: (Evaluation of environmental-influences)	Apisitpuvakul et al. (2008)
Japan	TV-sets, refrigerators, and another home-electric-device: (Investigation of alternative life-cycle-strategies)	Nakamura and Kondo (2006)
Korea	Personal-computers: (Investigation of the life-cycle environmental-influences)	Choi et al. (2006)

Zaman (2013) has explored the key features of emerging waste management technologies as well as their SWOT analysis in his study, which are illustrated in Tables 1.23 and 1.24, respectively.

As illustrated in Table 1.24, the present technologies have been analyzed based on SWOT analysis in the context of potential strengths, weaknesses, opportunities, and threats.

TABLE 1.23 The Key-Features of Emerging Technologies in Waste Management

Process Type	Key-Features	Wastes-Handling Type	References
Dry-composting	• Dried-materials through dry-composting can be effectively extracted via anaerobic-digestions. • It could be a potential technology among other different composting such as in-vessel or tunnel composting, vermin-composting, and windrow-composting.	Organic-wastes, garden-wastes, biodegradable-wastes.	Demirci et al. (2005); Prabha et al. (2007); Avfall Sverige (2009); Walker et al. (2009)
Anaerobic-digestion (AD)	A biological-conversion of wastes is anaerobic-digestion, where three different stages occur such as: (a) hydrolysis; (b) Acidifications; and (c) methanizations, respectively.	Organic-wastes, food-wastes.	Alternative Resources (2006); MWIN-RCA (2006); Visvanathan (2006)
Sanitary-landfill	A biological waste-treatment technology with landfill-facility control is sanitary-landfill, where the artificial-liners are used to prevent leachate-pollutions as well as air-emissions.	MSW	Tchobanoglous and Kreith (2002); Ludwing et al. (2003); FCM (2004)
Pyrolysis-thermal processes	A thermal process of MSW treatment technology is pyrolysis, where the un-sorted MSW can be treated at 600–650°C in the absence of oxygen.	MSW	Finnveden et al. (2000); Halton (2007)
Gasification	• A thermal waste treatment technology is gasification, which can be briquetting, fermentations, fluidized-beds or thermal-cracking. This is done in a controlled-environment with limited-access to air. • Thermo-chemical biomass-gasification can be possible for both dry as well as wet biomass to produce synthesis-gas, methane-rich-gases, and hydrogen.	MSW	LEE (2001); Wile'n et al. (2004); Kruse (2008)
Plasma-arc	• A plasma-reactor is used in this system which houses one or more plasma-arc torches generating an extremely high temperature environment between 5,000 and 14,000°C by the application of high-voltage between two electrodes. • After scrubbing, the gas output comprise of mainly CO and H_2, and the liquefied produce is mostly methanol.	MSW	GOI (2001); Circeo (2009)

TABLE 1.23 *(Continued)*

Process Type	Key-Features	Wastes-Handling Type	References
Plasma-arc-gasification	• For plasma-gasification technology, the temperatures of reactor ranges from approximately 800°F for a cracking-technology to as high as 8,000°F. • The organic-fraction of the MSW is converted to gas composed of hydrogen, carbon monoxide, and carbon dioxide.	MSW	Alternative Resources (2006); Circeo (2009)
Pyrolysis-gasification	• A hybrid waste treatment technology is pyrolysis-gasification, where there occurs a net reduction in the emissions of the sulfur-dioxide and particulates. • The emission of oxides of nitrogen, and dioxins may remain similar to the other thermal waste treatment technology.	MSW	DEFRA (2004); Malkow (2004); Alternative Resources (2007); Cherubini et al. (2008)
Bio-chemical conversions, anaerobic-process	• Materials recovery, biological treatment, and subsequent fuel-preparation. • Predigestion-stage of heating to 70°C for 1 hour with mesophilic-digestion at 35°C, or a thermophilic-digestion process operating the whole digester at 57°C.	MSW	Greater London Authority (2003)
Hydrolysis	• For MSW, the oxynol-hydrolysis is not yet in commercial operation. • The four major processes involved are: (a) waste-preparation; (b) acid-hydrolysis; (c) fermentation; and (d) distillation.	MSW, Sewage-sludge	Biffa (2003); Alternative Resources (2006)
Hydro-pulping	• This method has been developed in order to hydro-pulp the wastes and recover paper-fibers from refuses.	Paper-wastes	GOI (2001)
Bioreactor-technology	• Wastes are processed to maximize the landfill-gas preparation. • The anoxic-stage is followed by the oxidation-phase, methane-formations, and nitrogen-concentration increases along with concentration of carbon-dioxide originating from methane-oxidation.	Organic-wastes	Ludwing et al. (2003)

TABLE 1.23 *(Continued)*

Process Type	Key-Features	Wastes-Handling Type	References
	• The combination of both mechanical and biological process occur with the primary aim at stabilizing the biologically degradable components.		
Solid wastes' conversion to proteins	• It was found in the laboratory investigations at "Louisana State University, USA," of the possibilities under aerobic-conditions, to convert the insoluble-celluloses contained in municipal wastes by cellulolytic-bacteria. • In order to use as proteins, the bacteria are further harvested from the medium. • It was found of a content of 50 to 60% crude-protein in a single produced cell-protein.	Cellulosic-wastes	GOI (2001)

TABLE 1.24 SWOT Analysis of the Emerging Technologies in Waste Treatment

Method	Strength	Weakness	Opportunity	Threats
Dry-composting	• Biological-process in an open or confined area. • Possibilities of getting nutrient-rich organic fertilizers and soil-conditioner from wastes. • Future preservation of dried-wastes.	• Biodegradable-wastes are only managed by this process. • Difficulties in emission-control from the system.	• Resource recovery and bio-fertilizer making opportunity. • Possibility of biogas generation from the dry wastes. • Possibility of recovery of higher volume of biogases from the bio-reactor. • Opportunity of ethanol-productions. • Opportunity of nutrient recovery from identified wastes. • Resource-recovery.	• Potential environmental-threats from emissions. • Threat of environmental-impacts and pollution from emissions. • Potential environmental-threat of emissions to water and the atmosphere. • Water-contaminations. • Threat of environmental pollution from used chemicals. • Potential threat to water and soil contamination if poorly managed. • Threat of environmental-degradations by the emissions to atmosphere.
Anaerobic-digestions (AD)	• Biochemical-process with facilities for energy recovery. • Final-residues can be used as fertilizers.	• Organic waste only can be managed. • Requirement of higher investment-costs.	• Opportunity of retrieving biogases, fuels, and manures from the AD-facilities.	• Potential threat of environmental-emissions.

TABLE 1.24 *(Continued)*

Method	Strength	Weakness	Opportunity	Threats
Sanitary-landfills	• A natural decomposition-process capable of handling different kinds of wastes of larger-volumes. • Management of wastes in a controlled environment.	• Requirement of huge land-area and complex as well as costly emissions-control. • Requirement of longer-time in reclaiming the restoration of landfill-lands.	• Opportunity of recovering of biogases from landfills. • Opportunity of managing wastes in more environment-friendly manner, if sanitary-landfill remains fully-functional.	• Potential environmental threats of soil, air, and water contaminations as a consequence of weak liners and poorer management-systems.
Pyrolysis-thermal processes	• Treatment of different wastes categories with lower volumes of final-residues.	• Higher investment-costs and inappropriate-technology for MSW.	• Opportunity of recovering resources and energy.	• Potential environmental-threat from emissions.
Gasification	• Capability to treat almost all types of wastes. • Lower final-residues are generated by this process.	• Higher investment-costs and the technology is still developing for MSW.	• Opportunity of recovering energy and heat from the gasification of MSW.	• Environmental-impacts through emission to the atmosphere.
Plasma-arc	• Capability to treat almost all types of wastes with lower disposable-residues.	• Newer technology for MSW management and higher investment-costs.	• Opportunity of recovering energy and heat.	• Environmental-impacts through emission to the atmosphere.
Pyrolysis-Gasification	• Hybrid thermal-process with capabilities of larger volumes of different wastes treatment.	• Emerging-technology with higher investment-costs.	• Opportunity of recovering resources and energy.	• Potential environmental-threats of water and air pollution.
Biochemical-conversions, anaerobic-process	• An integrated waste treatment process having both mechanical as well as biological treatments.	• Capacity of limited wastes treatment. • By the use of this technology, organic-wastes can be treated.	• Possibilities of energy and resource recovery.	• Potential environmental-threats of emissions to water and the atmosphere.

TABLE 1.24 *(Continued)*

Method	Strength	Weakness	Opportunity	Threats
Hydrolysis	• Chemical-processes of fruits or foods waste to ethanol-productions.	• Latest-technology having limited problem-solving capacity.	• Opportunity of ethanol-productions.	• Contamination of water.
Hydro-pulping	• Resource-recovery as well as reuse in pulp and paper industries.	• Only paper-wastes are managed by this process.	• Resource-recovery.	• Threat of environmental-pollution from used chemicals.
Bioreactor-technology	• Management of higher waste volumes as compared to traditional-landfills.	• Requirement of pre-processing of wastes.	• Recovery of higher volume of biogases from the bio-reactors.	• Environmental-threats owing to emissions from the technology.
Conversion of solid-wastes to proteins	• Wastes to nutrient conversions.	• At experimental-stage with lower problem-solving potential.	• Opportunity of nutrient recovery from wastes.	• No threats have been identified.

1.3 CHALLENGES AND OPPORTUNITIES IN WASTE MANAGEMENT

A number of and varieties of waste related challenges are posed to governments as well as communities. Especially the amounts of waste in developing countries are generated and increasing. As much of the wastes are poorly managed with improper collection, inadequate disposal sites, or contaminated waste with hazardous materials, this waste has created major impacts on human health, particularly for those living near disposal-sites. Wastes also have a varied range of environmental-impacts on air, water, and land. Moreover, the inefficient use of scarce resources results in abandoned and materials discarded as wastes that represent a higher economic as well as environmental cost to be borne by the society as a whole. The proper management of waste is not only a challenge, but also an emerging-opportunity throughout the world. A proper waste management provides an opportunity not only in avoiding the harmful impacts associated with wastes, but also in recovering the useful resources, realizing the social, environmental, as well as economic benefits and consequently to take appropriate steps on the way to a sustainable prospect. By separating these wastes at sources, the un-contaminated organic-fractions can be composted and/or anaerobically digested. The improvements in waste management bring simultaneous benefits but requiring lesser investments, delivering livelihoods and jobs, contributing to economic growth, protecting, and improving the public health and the environment, respectively. More progress can be achieved through the re-evaluation of production and consumption processes, so that all the adverse impacts, inefficiencies, and losses associated with the generation and management of wastes are reduced. An improved waste management offers benefits to the socially-marginalized sectors. As a vital role is always played by the informal-sectors in many developing economies, this it can be recognized, professionalized, protected, and integrated into the system of waste management. There is a requirement of governance for the waste management to take into account both the inter-relationships and complexities within as well as outside government. It is a cooperative process that requires the involvement of different interests, including government at local, regional, and national levels. Moreover, it is a challenge as well as an opportunity for the governance system to recognize all these interests, and for reconciling different perspectives.

1.3.1 GROWTH AND CHALLENGES TO THE INTERNATIONAL WASTE-MANAGEMENT SYSTEMS

A rapid urbanization and economic development have resulted in considerable-improvements in the well-being of a larger-fractions of the worldwide population in the last few decades. At the same time, an augmented as well as intensive resource-consumption has been occurred with a consequent environmental release of larger wastes amount (Blanchard, 1992; Gerbens-Leenes et al., 2010; Wenheng and Shuwen, 2008). According to the American National Academy of Sciences, within months an amount of 94% of the substances of the earth enter the waste-streams (Bylinsky, 1995). The end-products or by-products of the production as well as utilization process has been regarded as wastes (European Union, 2008).

While consuming the resources in the current linear model, the resources are processed that enter into the human-environment, transformed, utilized, and discarded to the environment in the form of solid, liquid, or gaseous form of wastes. During the last few decades, there has been an operational and technological development in waste management in order to resolve contemporary technical, environmental, and economic challenges. The current global waste management practices have been focused on reducing the impacts rather than preventing them, and instead of long-term sustainable measures, the "end-of-pipe" solutions are suggested for waste associated problems (Seadon, 2010). A number of initiatives, such as design-for-environment, cleaner-production, extended-producer-responsibility, and industrial-symbiosis were introduced in production as well as consumption system (Frosch and Gallopoulos, 1989; Lifset, 1993; OECD, 2006). In order to achieve the required resource efficiency, such technology and operation-based innovations have broadened the waste issues discussions. A proper accounting as well as control of all kinds of gaseous emissions and solid and liquid wastes is a basic requirement for an integrated approach to wastes management (Stiles, 1996). The issues related to wastes have been recognized as a global problem rather than local-problem in view of adverse environmental-impacts because of a significant amount of wastes emission (UNEP, 2011; UNFCCC, 2005).

The generation of waste is expected during the production and consumption of resources on account of the laws of thermodynamics. An additional un-intended outputs harmful to the environment are resulted because of the production and consumption of intentional goods (Kronenberg and Winkler, 2009). Consequently, during the extractions, productions, consumptions, and final treatments of the resources result in the production of un-intended gaseous, liquidous, and solidious by-products or/and end-products from a life-cycle perspective. The process of waste minimization mainly encompass three elements: prevention and/or reduction of the waste generation at sources; quality improvement of the generated wastes, such as reducing the hazards, encouraging reuse, recycling, and revival (EEA, 2002; OECD, 1998). Thus, waste minimization has been a broader term involving waste-prevention in addition to treatment-measures (EEA, 2002). The waste preventions involve taking measures to reduce the quantity of wastes through reuse or extension of the life-span of products, the adverse impacts of the generated waste on human health and the environment, the content of harmful substances in products as well as the materials (European Union, 2008, 2012). It has become very challenging study regarding the prevention measurements (Fell et al., 2010; Gottberg et al., 2010; Sharp et al., 2010; Wilson et al., 2012a; Wilts, 2012; Zorpas and Lasaridi, 2013). The development of waste management system has been strongly linked with the contemporary-drivers and implications. Moreover, it has been highly dependent on social, economic, political, and environmental issues. There are considerable variations in the perceived drivers for waste management throughout the world. The strive for improved public-health in many developing countries is still a key-driver for waste-collection, while it is not so in most of Europe (Wilson, 2007). A significant improvement has been reported in many cities in the lower- and middle-income countries for the waste disposal technologies (Scheinberg et al., 2010; Wilson et al., 2012b, 2013). Wilson et al. (2012b) have revealed of more difficulty in regulating and implementation of new efficient as well as standardized waste-treatment systems by highly active informal-sectors for collection of wastes in the low- and middle-income countries. Indeed, as compared to the formal-sectors, there are twice as many people in the informal-sectors (World Bank, 2005). From the mutual cooperation by integrating the informal-sectors into formal waste management planning can result in gaining of significant benefits (Velis et al., 2012; Wilson et al., 2006, 2012b). Because of lack of accessible information/data on wastes, several difficulties are posed in the planning as well as implementation of a sustainable waste management systems (Bai and Sutanto, 2002; Jin et al., 2006).

Moreover, the role of legislations or regulations cannot be ignored in the development of waste management systems and the lacking of stringent-legislation often result in the reliance on low technological resolutions (Kumar et al., 2009). A range of legislations and regulatory-instruments exist in governmental policies in the area of waste management that are important in fostering an efficient resources management as well as in environment friendly treatment of wastes.

1.3.2 DIFFICULTIES IN THE GLOBAL RESOURCE-MANAGEMENT SYSTEMS

An inextricable-link exists between waste amounts and economic growth through the increasing amount of wastes throughout the world. As compared with high income countries, the waste generation rates are relatively lower per capita for the low- and middle-income countries. There is a need of calls for an environmentally sound approach to waste management to go beyond the safe disposal or recovery of generated wastes and addressing the root-causes of the problem by an attempt to change the unsustainable pattern of production as well as utilization (UN, 1993).

During the life-cycle stages, the material resources undergo several transformations, and the products that enter the consumption systems leave the system in the form of various wastes. Thus, in addition to these transformations, this also result in changes in chemical, physical, and biological properties of the material-resources (Gößling-Reisemann, 2011). An energy-intensive as well as costly recycling operations are required owing to the dilution of the resources to a critically low-level due to the increase in product-composition complexity by the manufacturing and consumption systems (Ayres, 1994; Ayres and Kneese, 1989). Provided that the availability of energy is limited, the recycling of materials beyond certain limits will be prevented by the environmental-cost of pure substances production. Thus, the utilization of many non-renewable resources become inherently un-sustainable and any attempt to close their loop will result in greater damages (Vesilind et al., 2007). The recycling of complex waste-streams are associated with technological challenges and thereby reducing the potential-utilities (Gößling-Reisemann, 2011; Rechberger and Brunner, 2001). Over the last few decades, there has been an increased product diversity in numbers as well as varieties. Consequently, the waste management systems receiving and dealing with the waste materials have also increased and likely to increase further. Therefore, higher challenges are occurred for the waste management systems in handling the wastes in a more sustainable way on account of poor-sorting or complicated processes of waste treatment (United Nations Environment Program, 2011). For sustainable consumption as well as disposal of resources the environmental awareness in society plays a vital role. The behavior of poor waste-sorting among people hampers directly the recycling activities. An economical as well as environmental inefficient recycling operations take place if the recyclable materials are mixed with another kind of wastes making other treatment options such as incineration or landfills more attractive. It has been observed that a strong concern among people for a cleaner environment and belief of learning, information, and awareness campaigns play a significant role as drivers to behavioral-changes (Mbeng et al., 2009). However, the challenge lies in the successful association between the behavior of peoples' waste disposal as well as global-concerns (MORI, 2002). The failure of policy decisions in order to achieve the intended goals are due to a number of factors that are responsible for the practical implementation as well as performances of various innovations in resource management. Consequently, the safer disposals towards a sustainable resource-management throughout the products' life-cycle chain is still a major concern (Wilts et al., 2011).

1.3.3 MULTI-CRITERIA DECISION-MAKING CONTEXT IN WASTE MANAGEMENT

It has been reported by the World Bank of about 1.3 billion-tons of MSWs generation in 2011, which is estimated to increase by the year 2025 to 2.2 billion-tons (Hoornweg and Bhada-Tata, 2012). D-Waste (2014) has reported of MSWs generation of 1.84 billion-tons in 2013, and similarly, the United Nations Environment Program (2015) has estimated the current of MSWs generation to be approximately 2 billion-tons per-annum. With the diffusion of electronic and plastic products, people are discarding larger quantity of wastes having more complex composition than ever before (Herva et al., 2014; Vergara and Tchobanoglous, 2012). Thus, the field of solid waste management has attracted more concerns with the increase in urbanization, population, and economic-development (Levis et al., 2013). Apart from these elements, the lifestyles, income-levels, and socio-economic as well as cultural factors are the key drivers for patterns of waste generation and the waste-toxicity (Eriksson and Bisaillon, 2011; International Solid Waste Association, 2002). Therefore, in the 21st century, solid waste management has become the most challenging service-sectors for municipal-authorities (Cherubini et al., 2009; Zaman, 2014). Severe public health-risks, local, and global environment-hazards, and socio-economic problems are posed with regard to the solid waste issues like emissions in soil, air, and water, and soil (Ikhlayel et al., 2013). Through supportive tools in order to assess the overall performance of systems, these complex management requirements can be better controlled, which includes legal, administrative, economic as well as planning-aspects (Mendes et al., 2013). Eriksson et al. (2003) have revealed the major benefits of waste management models as their capability in dealing with complexities and uncertainties to handle different goals. The formalized assessment tools help in ensuring a structured-approach and in providing comprehensive methods to collect and analyze data (Zurbrügg et al., 2014). Furthermore, these models must have the ability to complement the present insights and also in adapting the rising complexities of current solid waste management (Manfredi and Goralczyk, 2013). However, since the late 1960s the decision support models in the waste management field have been developed (Karmperis et al., 2013), and during the 1980s and 1990s, for the assessment of entire waste management systems, some models were designed (Allesch and Brunner, 2014). A huge number of methods in the present context of waste management decisions result in facing the challenges in the choice of assessment methodology (Finnveden et al., 2007; Pires et al., 2011). The most widely used decision support methods in the field of waste management have been reported as "life-cycle assessment (LCA)," "cost-benefit analysis," and "multicriteria decision making (MCDM)" (Karmperis et al., 2013; Milutinović et al., 2014; Morrissey and Browne, 2004). By simultaneously applying multiple conflicting-criteria, the MCDM can effectively guide the decision-makers for evaluating the existing or potential alternatives (Belton and Stewart, 2002; Kou et al., 2011; Zhou et al., 2010). In the case of solid waste management, the MCDM methods are considered to be the most effective decision support tools in decision-making because of their abilities in handling several criteria (Soltani et al., 2015). Moreover, the choice of steps involved in the MCDM process such as selection of indicators, weighting, normalizing, and sensitivity analysis are critical aspect in affecting the results directly (Çelen, 2014; Dobbie and Dail, 2013; Dodgson et al., 2009; Ebert and Welsch, 2004).

The selection of the procedure in each MCDM step depends on the study purpose and the available information (Zardari et al., 2015).

1.4 CONCLUSION

The inextricable-linkage between financial-growth, consumption of resources and wastes has been a vital confront throughout the globe. Moreover, the waste management situation in the middle as well as low-income countries has become more severe because of lacking in infrastructures and rapid-growth in waste-amounts. Despite of relatively developed systems for waste management, the seriousness of waste management-related issues is rather more due to higher consumption level. Although, there has been a significant progress in waste management technologies, but there occurs significant barriers in sustainable resource-recovery-operations. The failure of different system interventions are due to the result of lack of recognition of various cross-level as well as multi-scale system dynamics.

KEYWORDS

- **decision-making**
- **multicriteria decision-making**
- **recycling**
- **resource-management**
- **waste generations**
- **waste management**

REFERENCES

Allesch, A., & Brunner, P. H., (2014). Assessment methods for solid waste management: A literature review. *Waste Management and Research, 32,* 461–473.

Alternative Resources, Inc., (2006). *Focused Verification and Validation of Advanced Solid Waste Management Conversion Technologies: Phase 2*. New York. Department of Sanitation.

Apisitpuvakul, W., et al., (2008). LCA of spent fluorescent lamps in Thailand at various rates of recycling. *Journal of Cleaner Production, 16*(10), 1046–1061.
ATSDR, (1992). *Agency for Toxic Substances and Disease Registry*. Case studies in Environmental medicine-lead toxicity. Atlanta: U. S. Department of Health and Human Services, Public Health Service.
ATSDR, (1999). *Agency for Toxic Substances and Disease Registry*. Toxicological profile for lead. Atlanta: U. S. Department of Health and Human Services, Public Health Service.
Avfall Sverige, (2009). RAPPORT U2009:07, Torrkonservering av matavfall från hushåll.
Ayres, U. R., & Kneese, A. V., (1989). Externalities: Economics and thermodynamics. *Economy and Ecology: Towards Sustainable Development*. Netherlands: Kluwer Academic Publishers.
Ayres, U. R., (1994). Industrial metabolism: Theory and policy. In: Ayres, R. U., & Simonis, U. E., (eds.), *Industrial Metabolism; Restructuring for Sustainable Development* (pp. 3–20). Tokyo: United Nations University Press.
Bai, R., & Sutanto, M., (2002). The practice and challenges of solid waste management in Singapore. *Waste Management, 22,* 557–567.
Barba-Gutiérrez, Y., Adenso-Diaz, B., & Hopp, M., (2008). An analysis of some environmental consequences of European electrical and electronic waste regulation. *Resources, Conservation, and Recycling, 52*(3), 481–495.
Bellinger, D. C., (1995). Interpreting the literature on lead and child development: The neglected role of the experimental system. *Neurotoxicology and Teratology, 17,* 201–212.
Belton, V., & Stewart, T., (2002). *Multiple Criteria Decision Analysis: An Integrated Approach* (p. 372). Berlin, Germany: Springer.
Biffa, (2003). *Thermal Methods of Municipal Waste Treatment*. UK.
Blanchard, O., (1992). Energy consumption and modes of industrialization: Four developing countries. *Energy Policy, 20,* 1174–1185.
Bulkeley, H., & Gregson, N., (2009). Crossing the threshold: Municipal waste policy and household waste generation. *Environment and Planning A, 41,* 929–945.
Bylinsky, G., (1995). Manufacturing for reuse. *Fortune, 131,* 102, 103.
Cardis, E., Vrijheid, M., Blettner, M., Gilbert, E., et al., (2005). Risk of cancer after low doses of ionizing radiation: Retrospective cohort study in 15 countries. *British Medical Journal, 331,* 77.
Celen, A., (2014). Comparative analysis of normalization procedures in TOPSIS method: With an application to Turkish deposit banking market. *Informatica, 25,* 185–208.
Central Pollution Control Board, (2011). *Status Report on Municipal Solid Waste Management*. Accessible at: cpcb.nic.in.
CETESB, (1991). Companhia de Tecnologia de Saneamento Ambiental. Séries Manuais. Avaliação de Desempenho de Estações de Tratamento de Esgotos (p. 37). São Paulo.
CETESB, (1993). Companhia de Tecnologia de Saneamento Ambiental. *Método de Ensaio-Salmonella: Isolamento e Identificação, Norma L5, 218.*
Cherubini, F., Bargigli, S., & Ulgiati, S., (2009). Life cycle assessment (LCA) of waste management strategies: Landfilling, sorting plant and incineration. *Energy, 34*(12), 2116–2123.
Cherubini, F., Bargigli, S., et al., (2008). Life cycle assessment of urban waste management: Energy performances and environmental impacts. The case of Rome, Italy. *Waste Manag. (Oxford), 28*(2008), 2552–2564.

Choi, B. C., et al., (2006). Life cycle assessment of a personal computer and its effective recycling rate. *The International Journal of Life Cycle Assessment, 11*(2), 7, 122–128.
Circeo, L. J., (2009). *Plasma Arc Gasification of Municipal Solid Waste*. Plasma applications research program.
CONAMA, (2006). Resolução n. 357-2006. Define critérios e procedimentos, para o uso agrícola de lodos de esgoto gerados em estações de tratamento de esgoto sanitário e seus produtos derivados, e dá outras providências. Brazil.
Corbett, S. J., Rubin, G. L., Curry, G. K., & Kleinbaum, D. G., (1993). The Sydney beach users' advisory group. The health effects of swimming at Sydney beaches. *American Journal of Public Health, 83,* 1701–1706.
Cox, J., Giorgi, S., Sharp, V., et al., (2010). Household waste prevention: A review of evidence. *Waste Management and Research, 28,* 193–219.
DEFRA, (2004). *Review of Environmental and Health Effects of Waste Management: Municipal Solid Waste and Similar Wastes*. F. A. R. A. Department for Environment. London, Enviros Consulting Ltd. and University of Birmingham with Risk and Policy Analysts Ltd., Open University and Maggie Thurgood.
Demirci, A., Cekmecelioglu, D., et al., (2005). Applicability of optimized in-vessel food waste composting for windrow systems. *Biosyst. Eng., 91*(4), 479–486.
Department of Fertilizers, (2017). *34th Report of the Standing Committee on Chemicals and Fertilizers on Implementation of Policy on Promotion of City Compost*. 16th Lok Sabha, Lok Sabha Secretariat, Government of India.
Dobbie, M. J., & Dail, D., (2013). Robustness and sensitivity of weighting and aggregation in constructing composite indices. *Ecological Indicators, 29,* 270–277.
Dodgson, J. S., Spackman, M., Pearman, A., et al., (2009). *Multi-Criteria Analysis: A Manual*. London, UK: Department for Communities and Local Government.
Duan, H., et al., (2009). Life cycle assessment study of a Chinese desktop personal computer. *Science of The Total Environment, 407*(5), 1755–1764.
D-Waste, (2014). *Waste Atlas 2013 Report* (p. 44). D-WASTE.
Ebert, U., & Welsch, H., (2004). Meaningful environmental indices: A social choice approach. *Journal of Environmental Economics and Management, 47,* 270–283.
EEA, (2002). Case studies on waste minimization practices in Europe. In: Jacobsen, H., & Kristoffersen M., (eds.), *Copenhagen: European Environment Agency.* Available at: http://www.eea.europa.eu/publications/topic_report_2002_2 (accessed 22 November 2019).
Eleutério, L., (1977). Diagnóstico da situação ambiental da cabeceira da Bacia do rio Doce, MG, no âmbito das contaminações por metais pesados em sedimentos de fundo [Dissertação]. Escola de Minas: Universidade Federal de Ouro Preto- Ouro Preto.
Elliott, P., Briggs, D., Morris, S., De Hoogh, C., Hurt, C., Jensen, T. K., Maitland, I., et al., (2001a). Risk of adverse birth outcomes in populations living near landfill sites. *British Medical Journal, 323*(7309), 363–368.
Elliott, P., Morris, S., Briggs, D., De Hoogh, C., Hurt, C., Jensen, T., Maitland, I., et al., (2001b). *Birth Outcomes and Selected Cancers in Populations Living Near Landfill Sites*. Report to the Department of Health, The Small Area Health Statistics Unit (SAHSU).
Environmental Defense – McDonald's Waste Reduction Task Force. (1991). McDonald's Corporation – Environmental Defense Waste Reduction Task Force. Retrieved from http://www.environmentaldefence.org/documents/927_McDonaldsfinalreport.htm (accessed on 18 February 2021).

Eriksson, O., & Bisaillon, M., (2011). Multiple system modeling of waste management. *Waste Management, 31,* 2620–2630.

Eriksson, O., Olofsson, M., & Ekvall, T., (2003). How model-based systems analysis can be improved for waste management planning. *Waste Management and Research, 21,* 488–500.

European Union, (2008). *European Union Waste Framework Directive*. European Union. Retrieved from: https://heinonline.org/HOL/Page?collection=journals&handle=hein.journals/colenvlp12&id=601&men_tab=srchresults (accessed on 18 February 2021).

European Union, (2012). *Guidelines on Waste Prevention Programs*. EU. Available at: http://ec.europa.eu/environment/waste/prevention/pdf/Waste%20Prevention_Handbook.pdf (accessed on 18 February 2021).

Evans, D., (2011). Blaming the consumer-once again: The social and material contexts of everyday food waste practices in some English households. *Critical Public Health, 21,* 429–440.

Evans, D., (2012). Beyond the throwaway society: Ordinary domestic practice and a sociological approach to household food waste. *Sociology, 46,* 41–56.

FCM, (2004). *Solid Waste as a Resource, Review of Waste Technologies*. Canada: 111.

Fell, D., Cox, J., & Wilson, D. C., (2010). Future waste growth, modeling and decoupling. *Waste Management and Research, 28,* 281–286.

Fielder, H. M. P., Poon-King, C. M., Palmer, S. R., Moss, N., & Coleman, G., (2000). Assessment of impact on health of residents living near the Nant-y-Gwyddon landfill site: Retrospective analysis. *British Medical Journal, 320,* 19–23.

Finnveden, G., Björklund, A., Moberg, A., et al., (2007). Environmental and economic assessment methods for waste management decision-support: Possibilities and limitations. *Waste Management and Research, 25,* 263–269.

Finnveden, G., Johansson, J., et al., (2000). Life cycle assessments of energy from solid waste. *Future-Oriented Life Cycle Assessments of Energy from Solid Waste*. Stockholm, Sweden.

Frosch, R. A., & Gallopoulos, N., (1989). Strategies for manufacturing. *Scientific American, 261,* 144–152.

Gerbens-Leenes, P. W., Nonhebel, S., & Krol, M. S., (2010). Food consumption patterns and economic growth. Increasing affluence and the use of natural resources. *Appetite, 55,* 597–608.

Giusti, L., (2009). A review of waste management practices and their impact on human health. *Waste Management, 29,* 2227–2239.

GOI, (2001). *Solid Waste Management Manuals from Government of India*. Ministry of Urban Development Government of India.

Gößling-Reisemann, S., (2011). Entropy production and resource consumption in life-cycle assessments. In: Bakshi, B. R., Gutowski, T. G., & Sekulic D. P., (eds.), *Thermodynamics and the Destruction of Resources*. Cambridge: Cambridge University Press.

Gottberg, A., Longhurst, P. J., & Cook, M. B., (2010). Exploring the potential of product-service systems to achieve household waste prevention on new housing developments in the UK. *Waste Management and Research, 28,* 228–235.

Government of India, (1995). *Ministry of Environment and Forests Gazette Notification No. 460 Dated* (pp. 10–20). New Delhi.

Goyer, R. A., (1996). Toxic effects of metals. In: Klaassen, C. D., (ed.), *Casarett and Doull's Toxicology: The Basic Science of Poisons* (pp. 691–736). New York: McGraw Hill.

Graham-Rowe, E., Jessop, D. C., & Sparks, P., (2014). Identifying motivations and barriers to minimizing household food waste. *Resources, Conservation and Recycling, 84,* 15–23.

Greater London Authority, (2003). *City Solutions: New and Emerging Technologies for Sustainable Waste Management (Final Report).* London.

Halifax Regional School Board, (2009). *Solid Waste Management Policy.*

Halton, (2007). *The Regional Municipality of Halton.* Step 1B: EFW Technology Overview. Halton EFW Business Case. Halton.

Herva, M., Neto, B., & Roca, E., (2014). Environmental assessment of the integrated municipal solid waste management system in Porto (Portugal). *Journal of Cleaner Production, 70,* 183–193.

Hoornweg, D., & Bhada-Tata, P., (2012). *What a Waste: A Global Review of Solid Waste Management.* Urban Development Series Knowledge Paper, No.15, Washington, USA: World Bank.

Ikhlayel, M., Higano, Y., Yabar, H., et al., (2013). Proposal for a sustainable and integrated municipal solid waste management system in Amman, Jordan, based on the life cycle assessment method. In: *Proceedings of the 23rd Pacific Conference of the Regional Science Association International (RSAI).* Bandung, Indonesia. Toyohashi, Japan: RSAI Pacific Section.

International Solid Waste Association, (2002). *Implementing the Three Dimensions of Sustainable Development. Industry as a Partner for Sustainable Development-Waste Management* (pp.17–23). Copenhagen, Denmark: ISWA.

IPCS, (1988). International program on chemical safety. *Environmental Health Criteria, 61-Chromium.* Geneva: World Health Organization (WHO) [Internet]. Available from: http://www.inchem.org/documents/ehc/ehc/ehc61.htm (accessed on 18 February 2021).

IPCS, (1992). International program on chemical safety. *Environmental Health Criteria, 134-Cadmium.* Geneva: World Health Organization (WHO) [Internet]. Available from: http://www.inchem.org/documents/ehc/ehc/ehc134.htm (accessed on 18 February 2021).

IPCS, (1995). International program on chemical safety. *Environmental Health Criteria, 165-Inorganic Lead.* Geneva: World Health Organization (WHO) [Internet]. Available from: http://www.inchem.org/documents/ehc/ehc/ehc165.htm (accessed on 18 February 2021).

IPCS, (2001). International program on chemical safety. *Environmental Health Criteria, 224-Arsenic and Arsenic Compounds.* Geneva: World Health Organization (WHO) [Internet]. Available from: http://www.inchem.org/documents/ehc/ehc/ehc224.htm (accessed on 18 February 2021).

Jin, J., Wang, Z., & Ran, S., (2006). Solid waste management in Macao: Practices and challenges. *Waste Management, 26,* 1045–1051.

Karmperis, A. C., Aravossis, K., Tatsiopoulos, I. P., et al., (2013). Decision support models for solid waste management: Review and game-theoretic approaches. *Waste Management, 33,* 1290–1301.

Kay, D., Fleisher, J. M., Salmon, R. L., Jones, F., Wyer, M. D., Godfree, A. F., Zelenauch-Jacquotte, Z., & Shore, R., (1994). Predicting likelihood of gastroenteritis from sea bathing: Results from randomized exposure. *Lancet, 344*(8927), 905–909.

Kiddee, P., Naidu, R., & Wong, M. H., (2013). Electronic waste management approaches: An overview. *Waste Management, 33*(5), 1237–1250.

Kim, J., et al., (2004). Methodology for recycling potential evaluation criterion of waste home appliances considering environmental and economic factor. In: *Electronics and the Environment, Conference Record 2004*. IEEE International Symposium on 2004, IEEE.
Kou, G., Miettinen, K., & Shi, Y., (2011). Multiple criteria decision making: Challenges and advancements. *Journal of Multi-Criteria Decision Analysis, 18,* 1–4.
Kronenberg, J., & Winkler, R., (2009). Wasted waste: An evolutionary perspective on industrial by-products. *Ecological Economics, 68,* 3026–3033.
Kruse, A., (2008). Review of hydrothermal biomass gasification. *J. Supercrit. Fluids, 47*(2009), 391–399.
Kumar, S., Bhattacharyya, J. K., Vaidya, A. N., Chakrabarti, T., Devotta, S., & Akolkar, A. B., (2009). Assessment of the status of municipal solid waste management in metro cities, state capitals, class I cities, and class II towns in India: An insight. *Waste Management, 29,* 883–895.
Lee, A. O., (2001). *Refuse Derived Briquette Gasification Process and Briquetting Press.* World Intellectual Property Organization (WO/2001/034732).
Levis, J. W., Barlaz, M. A., De Carolis, J. F., et al., (2013). A generalized multistage optimization modeling framework for life cycle assessment-based integrated solid waste management. *Environmental Modeling and Software, 50,* 51–65.
Lifset, R., (1993). Take it back: Extended producer responsibility as a form of incentive-based policy. *Journal of Resource Management and Technology, 21,* 163–175.
Lina, F., Ballesteros, J. A., Teixeira, S. I., & Mussatto, (2014). Chemical, functional, and structural properties of spent coffee grounds and coffee silver skin. *Food Bioprocess Technol, 7,* 3493–3503.
Lu, L. T., et al., (2006). Balancing the life cycle impacts of notebook computers: Taiwan's experience. *Resources, Conservation and Recycling, 48*(1), 13–25.
Ludwing, C. S., Hellweg, et al., (2003). Municipal solid waste management; strategies and technologies for sustainable solutions. *Int. J. Life Cycle Assess, 8,* 2–114.
Malkow, T., (2004). Novel and innovative pyrolysis and gasification technologies for energy-efficient and environmentally sound MSW disposal. *Waste Manage (Oxford), 24,* 53–79.
Manfredi, S., & Goralczyk, M., (2013). Life cycle indicators for monitoring the environmental performance of European waste management. *Resources, Conservation and Recycling, 81,* 8–16.
Mavropoulos, A., et al., (2012). *Phase 1: Concepts and Facts, Globalization and Waste Management.* International Solid Waste Association.
Mayers, C. K., France, C. M., & Cowell, S. J., (2005). Extended producer responsibility for waste electronics: An example of printer recycling in the United Kingdom. *Journal of Industrial Ecology, 9*(3), 169–189.
Mbeng, L., Probert, J., Phillips, P., & Fairweather, R., (2009). Assessing public attitudes and behavior to household waste management in Cameroon to drive strategy development: A Q methodological approach. *Sustainability, 1,* 556–572.
Mendes, P., Santos, A. C., Nunes, L. M., et al., (2013). Evaluating municipal solid waste management performance in regions with strong seasonal variability. *Ecological Indicators, 30,* 170–177.
Metro Vancouver, (2010). *Integrated Solid Waste and Resource Management.*

Milutinović, B., Stefanović, G., Dassisti, M., et al., (2014). Multicriteria analysis as a tool for sustainability assessment of a waste management model. *Energy, 74,* 190–201.

MORI, (2002). *Public Attitudes Towards Recycling and Waste Management.* MORI Social Research Institute.

Morrissey, A. J., & Browne, J., (2004). Waste management models and their application to sustainable waste management. *Waste Management, 24,* 297–308.

MWIN-RCA, (2006). *Municipal Solid Waste (MSW) Options: Integrating Organics Management and Residual Treatment Disposal.* Workshop Report. Alberta.

Nakamura, S., & Kondo, Y., (2006). A waste input-output life-cycle cost analysis of the recycling of end-of-life electrical home appliances. *Ecological Economics, 57*(3), 494–506.

Nova Scotia Environment, (2009). *Solid Waste-Resource Management Strategy.* Retrieved from: http://www.gov.ns.ca/nse/waste/swrmstrategy.asp (accessed on 18 February 2021).

OECD, (1998). *Waste Minimization in OECD Member Countries.* OECD. Available at: http://search.oecd.org/officialdocuments/displaydocumentpdf/?doclanguage=en&cote=env/epoc/ppc(97)15/rev2 (accessed on 18 February 2021).

OECD, (2006). *EPR Policies and Product Design: Economic Theory and Selected Case Studies* [Online]. OECD. Available at: http://www.oecd.org/document/19/0,3746,en_2649_34281_35158227_1_1_1_1,00.html (accessed on 18 February 2021).

Paepatung, N., Nopharatana, A., & Songkasiri, W., (2009). Bio-methane potential of biological solid materials and agricultural wastes. *Asian J. Energy. Env., 10,* 19–27.

Parvathamma, G. I., (2014). An analytical study on problems and policies of solid waste management in India: Special reference to Bangalore city. *J. Environ. Sci. Toxicol. Food Technol., 8,* 6–15. doi: 10.9790/2402–081010615.

Phillips, P. S., Tudor, T., Bird, H., et al., (2011). A critical review of a key waste strategy initiative in England: Zero waste places projects 2008–2009. *Resources, Conservation and Recycling, 55,* 335–343.

Pires, A., Martinho, G., & Chang, N. B., (2011). Solid waste management in European countries: A review of systems analysis techniques. *Journal of Environmental Management, 92,* 1033–1050.

Planning Commission, Government of India, (2014). *Report of the Task Force on Waste to Energy in the Context of Integrated Municipal Solid Waste Management* (Vol. I). Available at: *http://planningcommission.nic.*in/*reports/genrep/rep_wte1205.pdf* (accessed on 18 February 2021).

Prabha, K. P., Loretta, Y. L., et al., (2007). An experimental study of vermi-biowaste composting for agricultural soil improvement. *Bioresour. Technol., 99*(2008), 1672–1681.

Quináglia, G. A., (2001). Estabelecimento de um protocolo analítico de preparação de amostras de solo para determinação de metais e sua aplicação em um estudo de caso [Dissertação]. São Paulo: Faculdade de Saúde Pública: Universidade de São Paulo.

Rechberger, H., & Brunner, P. H., (2001). A new, entropy-based method to support waste and resource management decisions. *Environmental Science and Technology, 36,* 809–816.

Rivas, B., Torrado, A., Torre, P., Converti, A., & Domínguez, J. M., (2008). Submerged citric acid fermentation on orange peel auto hydrolysate. *J. Agric. Food Chem., 56,* 2380–2387.

Rodrigues, R. B., Da Silva, J. T. A. F., & Maringoni, A. C., (2006). Efeito da aplicação de lodo de esgoto na severidade da murcha-de-curtobacterium em feijoeiro. *Summa Phytopathologica, 32,* 82–84.

Rudra, S. G., Nishad, J., Jakhar, N., & Kaur, C., (2015). Food industry waste: Mine of nutraceuticals. *Intern. J. Sci. Environ. Technol., 4*(1), 205–229.

Rushton, L., (2003). Health hazards and waste management. *British Medical Bulletin, 68,* 183–197. doi: 10.1093/bmb/ldg034.

Scheinberg, A., Wilson, D. C., & Rodic, L., (2010). *Solid Waste Management in the World's Cities: Water and Sanitation in the World's Cities 2010.* London, UK: Earthscan For Un-Habitat.

Seadon, J. K., (2010). Sustainable waste management systems. *Journal of Cleaner Production, 18,* 1639–1651.

Sharp, V., Giorgi, S., & Wilson, D. C., (2010). Methods to monitor and evaluate household waste prevention. *Waste Management and Research, 28,* 269–280.

Soltani, A., Hewage, K., Reza, B., et al., (2015). Multiple stakeholders in multicriteria decision-making in the context of municipal solid waste management: A review. *Waste Management, 35,* 318–328.

Stiles, S. C., (1996). A multi-media, integrated approach to environmental compliance and enforcement. *Industry and Environment, 19,* 12–14.

Tandon, H. L. S., (1997). Organic resources: An assessment of potential supplies, their contribution to agricultural productivity and policy issues for Indian agriculture from 2000–2025. In: Kanwar, J. S., & Katyal, J. C., (eds.), *Plant Nutrient Needs, Supply, Efficiency and Policy Issues, 2000–2025* (pp. 15–28). National Academy of Agricultural Sciences, New Delhi.

Tchobanoglous, G., & Kreith, F., (2002). *Handbook of Solid Waste Management* (2nd edn., p. 950). McGraw-Hill.

Thyberg, K. L., & Tonjes, D. J., (2015). A management framework for municipal solid waste systems and its application to food waste prevention. *Systems, 3,* 133–151.

Tomatis, L., (1990). *Cancer: Causes, Occurrence and Control.* International Agency for Research on Cancer, France.

U.S Environmental Protection Agency, (1995). *Decision-Makers' Guide to Solid Waste Management* (Vol. II). Washington, D.C. Retrieved from: http://www.epa.gov/osw/nonhaz/municipal/dmg2/ (accessed on 18 February 2021).

UN, (1993). *Report of the United Nations Conference on Environment and Development.* New York: United Nations.

UNEP, (2010). Division of Technology, Industry and Economics International Environmental Technology Center Osaka/Shiga. *Waste and Climate Change-Global Trends and Strategy Framework, UNEP.*

UNEP, (2011). *Towards a Green Economy: Pathways to Sustainable Development and Poverty Eradication: Part II.* Investing in Energy and Resource Efficiency.

UNFCCC, (2005). *Key GHG Data.* Greenhouse gas emissions data for 1990–2003 Submitted to the United Nations Framework Convention on Climate Change. UNFCCC Secretariat, Bonn, Germany. Available at: http://unfccc.int/resource/docs/publications/key_ghg.pdf (accessed on 18 February 2021).

United Nations Environment Program, (2007). *Montreal Protocol on Substances that Deplete the Ozone Layer 2007: A Success in the Making.* The United Nations Ozone Secretariat, United Nations Environment Program.

The United Nations Environment Program, (2015). Waste management: Global status. *Global Waste Management Outlook* (pp. 51–124). Osaka, Japan: UNEP.

The University of the Sunshine Coast, (2010). *Waste Management*. Retrieved from: https://www.usc.edu.au/about/policies-and-procedures (accessed on 18 February 2021).
The University of Victoria, (2004). *Waste, Recycling and Composting. Sustainability*. Retrieved from: http://web.uvic.ca/sustainability/WasteRecyclingComposting.htm (accessed on 18 February 2021).
USEPA, (1978). United states environmental protection agency. *Reviews of the Environmental Effects of Pollutants: III: Chromium* (p. 285). Washington.
Vega-Dienstmaier, J. M., Salinas-Piélago, J. E., Gutiérrez-Campos, M. R., Mandamiento-Ayquipa, R. D., Yara-Hokama, M. C., Ponce-Canchihuamán, J., & Castro-Morales, J., (2006). Lead levels and cognitive abilities in Peruvian children. *Revista Brasileira De Psiquiatria, 28,* 33–39.
Velis, C. A., Wilson, D. C., Rocca, O., Smith, S. R., Mavropoulos, A., & Cheeseman, C. R., (2012). An analytical framework and tool ('intera') for integrating the informal recycling sector in waste and resource management systems in developing countries. *Waste Management and Research,* 30, 43–66.
Vergara, S. E., & Tchobanoglous, G., (2012). Municipal solid waste and the environment: A global perspective. *Annual Review of Environment and Resources, 37,* 277–309.
Vesilind, P., Heine, L., & Hamill, S., (2007). Kermit's lament: It is not easy being green. *Journal of Professional Issues in Engineering Education and Practice, 133,* 285–290.
Viel, J. F., Arveux, P., Baverel, J., & Cahn, J. Y., (2000). Soft-tissue sarcoma and non-Hodgkin's lymphoma clusters around a municipal solid waste incinerator with high dioxin emission levels. *American Journal of Epidemiology, 152,* 13–19.
Viel, J. F., Daniau, C., Goria, S., Fabre, P., De Crouy-Chanel, P., Sauleau, E. A., & Empereur-Bissonnet, P., (2008). Risk of Hodgkin's lymphoma in the vicinity of French municipal solid waste incinerators. *Environmental Health,* 7(29), 51. http://www.ehjournal.net/content/7/1/51 (accessed on 18 February 2021).
Visvanathan, C., (2006). *Anaerobic Digestion of Municipal Solid Waste in Asia*. Asian Institute of Technology, Thailand. Tchobanoglous, G., & Kreith, F., (2002). *Handbook of Solid Waste Management*. McGraw-Hill, New York.
Walker, L., Charles, W., et al., (2009). Comparison of static, in-vessel composting of MSW with thermophilic anaerobic digestion and combinations of the two processes. *Bioresour. Technol.*
WCRF., & AICR, (1997). *Nutrition and the Prevention of Cancer: A Global Perspective*. World Cancer Research Fund (WCRF) and American Institute for Cancer Research (AICR), Washington, DC, USA.
Wenheng, W., & Shuwen, N., (2008). Impact study on human activity to the resource environment based on the consumption level difference of China's Provinces or autonomous regions. *China Population, Resources and Environment, 18,* 121–127.
Weshahy, A. A., & Rao, V. A., (2012). Potato peel as a source of important phytochemical antioxidant nutraceuticals and their role in human health: A review. *Phytochemicals as Nutraceuticals-Global Approaches to Their Role in Nutrition and Health,* 207–224.
Wile'n, C., Salokoski, P., et al., (2004). *Finnish Expert Report on Best Available Techniques in Energy Production from Solid Recovered Fuels*. Finish Environment Institute, Helsinki.
Williams, I. D., Schneider, F., & Syversen, F., (2015). The "food waste challenge" can be solved. *Waste Management, 41,* 1, 2.

Wilson, D. C., (2007). Development drivers for waste management. *Waste Management and Research, 25*, 198–207.
Wilson, D. C., Parker, D., Cox, J., et al., (2012a). Business waste prevention: A review of the evidence. *Waste Management and Research, 30,* 17–28.
Wilson, D. C., Rodic, L., Scheinberg, A., Velis, C. A., & Alabaster, G., (2012b). Comparative analysis of solid waste management in 20 cities. *Waste Management and Research, 30,* 237–254.
Wilson, D. C., Velis, C. A., & Rodic, L., (2013). Integrated sustainable waste management in developing countries. In: *Proceedings of the Ice-Waste and Resource Management* (Vol. 166, pp. 52–68). Available at: https://eprints.whiterose.ac.uk/78792/13/Wilson%20et%20al.1.pdf (accessed on 18 February 2021).
Wilson, D. C., Velis, C., & Cheeseman, C., (2006). Role of informal sector recycling in waste management in developing countries. *Habitat International, 30,* 797–808.
Wilts, H., (2012). National waste prevention programs: Indicators on progress and barriers. *Waste Management and Research, 30,* 29–35.
Wilts, H., Bringezu, S., Bleischwitz, R., Lucas, R., & Wittmer, D., (2011). Challenges of metal recycling and an international covenant as possible instrument of a globally extended producer responsibility. *Waste Management and Research, 29,* 902–910.
World Bank, (2005). *Waste Management in China: Issues and Recommendations.* Urban Development Working Papers No. 9, East Asia Infrastructure Department. The World Bank.
Zaman, A. U., (2013). Identification of waste management development drivers and potential emerging waste treatment technologies. *Int. J. Environ. Sci. Technol., 10,* 455–464. doi: 10.1007/s13762-013-0187-2.
Zaman, A. U., (2014). Measuring waste management performance using the 'zero waste Index': The case of Adelaide, Australia. *Journal of Cleaner Production, 66,* 407–419.
Zambon, P., Ricci, P., Bovo, E., Casula, A., Gattolin, M., Fiore, A. R., Chiosi, F., & Guzzinati, S., (2007). Sarcoma risk and dioxin emissions from incinerators and industrial plants: A population-based case-control study (Italy). *Environmental Health, 6,* 19.
Zardari, N. H., Ahmed, K., Shirazi, S. M., et al., (2015). *Weighting Methods and Their Effects on Multi-Criteria Decision Making Model Outcomes in Water Resources Management* (pp. 12–33). Berlin, Germany: Springer.
Zhou, P., Fan, L. W., & Zhou, D. Q., (2010). Data aggregation in constructing composite indicators: A perspective of information loss. *Expert Systems with Applications, 37,* 360–365.
Zorpas, A. A., & Lasaridi, K., (2013). Measuring waste prevention. *Waste Management, 33,* 1047–1056.
Zorpas, A. A., Lasaridi, K., Voukkali, I., et al., (2015). Household waste compositional analysis variation from insular communities in the framework of waste prevention strategy plans. *Waste Management, 38,* 3–11.
Zurbrügg, C., Caniato, M., & Vaccari, M., (2014). How assessment methods can support solid waste management in developing countries: A critical review. *Sustainability, 6,* 545–570.

CHAPTER 2

WASTE MANAGEMENT: AN OVERVIEW

ABSTRACT

Waste management has been a challenging issue in developing countries such as India for urban and rural developments. Each country has its socioeconomic-environmental considerations that might not accept some unique disposal methods as optimal choices. Under vagueness and incomplete information, selecting appropriate disposal methods of waste management is traditionally discussed as a discrete, "multi-criteria" decision-making problem. In this study, a comprehensive analysis of the works carried out was made about India's waste management strategies. The outcome will help decision-makers recommend or propose the preferable choices for waste disposal and treatment methodologies to prevent and protect people from the harmful effects of waste.

2.1 INTRODUCTION

"There are barely any things sure throughout everyday life-one is demise, second is change, and the other is waste." Though the wastes cannot be stopped to generate, but with appropriate management practices, these can be minimized. Any material which is not required by the owner, maker, or processor is regarded as waste. The generation of wastes is inescapable in every locale, howsoever large or little. Since the beginning of development, humankind has bit by bit strayed from nature, and today there has been an intense change in the way of life of the human culture. The direct impression of this change is found in the nature and amount of garbage that a community creates. The wastes can be properly disposed or reused through appropriate management. The Indian cities that are competing at a faster rate with worldwide economies in their drive for quick financial improvement have so far neglected to viably deal with the enormous amount of waste produced. The quantum of waste created in Indian towns and urban communities is expanding day-by-day by virtue of its expanded GDP and expanding population. The yearly amount of solid wastes produced in Indian urban communities has expanded from 6,000,000 tons in 1947 to 1948 million tons in 1997 with a yearly development pace of 4.25%, and by 2047 it is expected to have an increment to 300 million tons (CPCB, 1998). Population blast combined with improved way of individuals' life brings about expanded generations of wastes in Indian cities and villages. The activities of human-beings have consistently produced wastes, which was not a significant issue when the human-population was relatively lesser and migrant, however, this turned into a major issue with urbanization and the development of huge-cities. Poor-management of wastes have driven in tainting of water, soil, and air, and to a significant threat on general well-being. A portion of the immediate health-impacts by the mismanagement of wastes are outstanding and can be found particularly in developing countries. As science and innovation built up, the management of expanding volumes of wastes on a regular basis have turned into an exceptionally-organized, specific, and complex-action. The attributes of waste-materials advanced in accordance with the changing way of life, and the quantity of new synthetic-substances present in the different waste streams have been expanded significantly. It is difficult to measure the long-term health effects by exposing to substances present in wastes, or generated at waste disposal facilities, predominantly with little-concentrations and when other introduction-pathways occur. The motivation behind this study is to pick-up information about different activities in India, and to find the degree for development in the management of wastes.

2.1.1 CATEGORIZATION OF WASTES

In India, the generated wastes may be of different types such as: domestic wastes, industrial wastes, agricultural wastes, electronic wastes, paper wastes, bio-medical wastes, etc. Further, the wastes can be classified as follows: solid wastes including vegetable wastes, kitchen wastes, household wastes, etc.; e-wastes (EW) including discarded electronic devices; liquid wastes including utilization of water in different industries, thermal power plants, distilleries, etc.; plastic wastes including plastic bags, buckets, bottles, toys, etc.; metal wastes including un-used metal sheets, metallic scraps, etc.; and nuclear wastes including un-used materials of nuclear power-plants, etc. Moreover, these wastes can be grouped into two types, such as wet-wastes (bio-degradable) and dry-wastes (non-biodegradable). Wet-wastes (bio-degradable) include kitchen wastes, flowers, and fruits wastes, sanitary wastes, etc. While, dry-wastes (non-biodegradable) include paper and plastic, all kind of packaging, glasses, rubbers, dust from house sweeping, ashes, and discarded electronic items, furniture, and equipment from houses, offices, colonies, etc. Figures 2.1–2.3 illustrate the vegetables wastes, municipal solid wastes (MSWs), and plastic wastes, respectively.

FIGURE 2.1 Vegetables wastes.

2.2 LITERATURE

In India, there are over 23,000 primary health centers, about 6,00,000 hospital beds, thousands of registered nursing-homes, and limitless unregistered nursing homes as well as dispensaries. The hospitals are tertiary-care hospitals generally linked with teaching colleges, more than

FIGURE 2.2 Municipal solid wastes.

FIGURE 2.3 Plastic wastes.

2,000 district hospitals, and with healthcare dispensaries. The data of innumerable pathology laboratories is hardly accessible (Patil and Shekdar, 2001). An increased incidence of musculoskeletal problems and accidents has been found to be faced by waste management workers. There is a requirement of assessment and monitoring of the health impacts of new technologies in waste management, the increasing use of recycling, and composting, respectively (Rushton, 2003). A major source of hazardous chemicals as well as contaminants is waste (Lee and Jones-Lee, 1994), and proper management of wastes can be viewed as an element of the environmental management process covering the three pillars of sustainability such as social, economic, and political development (Pope et al., 2004). Around the world, the waste management division contributes roughly 3–5% of absolute anthropogenic outflow in 2005. Contrasted with the complete outflow, this rate is relative minor (Bogner et al., 2007). However, the waste division is in an express that it moves from being a minor wellspring of worldwide emanations to turning into a significant emissions saver (United Nations for Environmental Program, 2010). Through waste hierarchy principles, the reduction in emission from waste sectors can be achieved that include the least preferred option as the disposal for managing wastes, and avoiding and minimizing as the most preferred option (European Environment Agency, 2007). Liyala (2011) and Oberlin (2011) have revealed the poor policy and legislation, lacking in the political willingness, lacking in the public commitments, lacking in the technical capacities, and inadequate financing as the common causes for poor waste management services. While different authors consider it separately, for instance, as technical (Scheinberg, 2011), as political, economic or institution oriented (Wilson et al., 2010), as poor attitudes of the public towards waste collections and disposal or treatment (Liyala, 2011; Oberlin, 2011). The environmental management institutions get weakened by the political interference and create a community having difficulty in working (Okot-Okumu and Nyenje, 2011). Mbeng et al. (2009) have recommended for simple and accessible awareness programs to change the urban residents' mindset in perceiving wastes as resources instead of something without value. Zaman (2013) has identified key social, economic, and environmental drivers from case studies of waste management development in Sweden. The key social drivers included personal behaviors, local practices of waste management, consumption, and generation of wastes. The economic drivers included the waste resource value, financial benefits from facilities of waste treatment and landfill tax. The environmental drivers included the global changes in climate, environmental movements as well as awareness. Further, the potential emerging technologies for waste management systems were identified as dry composting, plasma arc, pyrolysis-gasification, and anaerobic digestion, respectively. The waste delivered in urban territories of India is roughly 1,70,000 tons for every day, equal to about 62 million tons for every year, and this is relied upon to increment by 5% every year owing to increase in population and changing ways of life (Planning Commission, Government of India, 2014). Urban India created 31.6 million tons of waste in 2001, and is producing 47.3 million tons at present. Waste generation has been predicted to be 161 million-tons by 2041, a five-fold increment in four-decades (Annepu, 2012). Chandigarh has been revealed as the first city for developing the solid waste management in an arranged manner and has improved waste administration in contrast to other Indian urban communities (Rana et al., 2015). The informal sectors have a significant job in India and this must be incorporated into formal solid waste management frameworks (Annepu, 2012). Usually, the informal sectors are characterized by small scale, largely unregulated, labor-intensive, un-registered little-technology or manufacturing or provisions of materials, and administrations (Wilson et al., 2006). The various wastes produced because of the progression of innovation may have a significant impact on the environment as well as the public, if not appropriately put-away, gathered, moved, treated, and discarded. Thus, it is becoming issues of concern, particularly in developing and transition countries around the globe, about e-waste generations, treatments, and disposal, to the professionals in waste management, innumerable NGOs and residents, and international agencies as well as governments. E-waste stream contains assorted materials, which requires extraordinary treatment and cannot be dumped in landfill destinations, most conspicuously, unsafe substances, for example, lead, polychlorinated biphenyls, mercury, polybrominated biphenyls, brominated flame retardants, poly-brominated diphenyl ethers, and important substances like steel, iron, aluminum, copper, silver, gold, platinum, plastics, and palladium (Chancerel et al., 2009; Widmer et al., 2005). The manufacturers are producing superior products with the rapid progress of technology, like televisions, the latest, and smarter mobile phones, and innovative computing devices at a rapid rate. As in order to fulfill the needs of people more and more electronic products are produced worldwide, thus these items are produced by using more resources. Hence, the production of electronic waste is also increasing in a rapid way for computing and other information as well as communication equipment. Moreover, the electronic products selling in countries like China, India, and across Latin America and Africa have been predicted to increase sharply in the next 10 years, which is a higher growth pattern and will be influenced not only by requirement but also by modifications in design, technology, and marketing initiatives (Huisman et al., 2007). During the provision of health care services by health care units, all wastes produced are referred to as health care wastes including infectious as well as non-infectious waste materials, hazardous wastes, chemicals, and other non-hazardous wastes. The bio-hazardous wastes are constituted of infectious wastes such as sharps, non-sharps, liquid waste, plastic disposables, etc., and non-infectious wastes such as radioactive waste, chemical waste, discarded glasses, cytotoxic wastes, and incinerated wastes. Toxic chemicals like mercury, formalin, and xylene are supposed to be present in wastes from hospitals or pathological laboratories (Manzurul et al., 2008). Though 75 to 90% of the hospital wastes are non-hazardous as well as harmless compared to other municipal wastes, the left-over percentage is hazardous to human being or animals, and damaging the environment (Dwivedi et al., 2009). The patients and personnel handling the hospital wastes are exposed to risks as it is infectious, hazardous, and creating serious threats to environmental health. Thus prior to its final disposal, it requires specific treatment along with management (Abdulla et al., 2008). For improving the usefulness and quality of epidemiological studies that has been applied to populations living in regions where waste management facilities are planned and located, preference should be given to prospective group studies of enough statistical power with respect to direct-human-exposure measurements, and data on health-effect as well as susceptibility biomarkers (Giusti, 2009). Inappropriate disposal practices of hospital wastes affect the people getting in direct-contact. A variety of disease vectors including mosquitoes and flies are attracted by waste piles (Dinesh et al., 2010) and can result in the growth of insects, un-pleasant odors, environmental pollution, and rodents and worms, leading in transmission of diseases like cholera, typhoid, tuberculosis, HIV, Hepatitis B and C by the injuries from sharps infected with human bloods (Manyele and Lyasenga, (2010). The beneficial microbes and bacteria in septic systems may be killed by the antibiotics poured down the drain, and by dumping the healthcare wastes in un-controlled areas can result in a direct environmental effect by contamination of soils and underground waters. Inappropriate filtration of flue gases during incineration may cause air pollution that result in the illnesses to the nearby population (Verma, 2010). An estimation of the "total waste electrical and electronic equipment (WEEE)" by 2020 indicated a growth between 2.5–2.7% annually, with an accomplishment of a total of approximately 12.3 million tons, because of the increase in the number of appliances entering the market every year in developed and developing nations (Schluep et al., 2009). Particularly a complex waste is encompassed in the current e-waste issues in terms of products varieties (ITU, 2011a, b).

India is shifting rapidly from a farm-oriented nation to services and industrial oriented nation. The urban areas are constituted by about 31.2% population with over 377 million urban people living in 7,935 towns or cities. In the three megacities, namely Delhi, Greater Mumbai, and Kolkata population is more than 10 million. There is a population of 100,000 or more in 415 cities and more than 1 million populations in 53 cities (Census, 2011). Moreover, the cities with more than 10 million populations are mostly State-capitals, Union-Territories, and other industry/business-oriented centers. The residents living in different zones in India having different geographic and climatic regions have different pattern of consumption and waste generation. However, no concrete steps had been taken till date to analyze the patterns of waste generation in regional and geographical context for the urban towns and limited data is available for researchers based on the study conducted by "National Engineering and Environmental Research Institute (NEERI), Nagpur," "Central Pollution Control Board (CPCB), New Delhi," "Federation of Indian Chambers of Commerce and Industry (FICCI, 2009), New Delhi," and "Central Institute of Plastics Engineering and Technology (CIPET), Chennai." Joshi and Ahmed (2016) have tried to evaluate the major parameters of municipal solid waste management (MSWM). They have concluded that in metropolitan cities/towns of developing countries like India, the de-centralized solid waste processing unit's installation and development in formal recycling industrial sectors is essential. With respect to cleaner energy as well as cleaner city, a major decision-making issue has been regarded as the cleaner waste management. Thus there is a specific requirement of integrated theoretical analytical-framework research on cleaner waste management in developing nations (Zheng et al., 2016). Alves et al. (2016) have made an attempt to evaluate the tools used and the efficiency of the adopted methods in order to define the proper way for waste management in the Rock in Rio-event, on events without disturbing the dynamics of cities and promoting social inclusions and waste valuation as well. A huge problem in many developing countries including India is mainly dealt with the disposal of bio-medical wastes, contaminated sharps, and common garbage. Though different waste management rules exists in India to manage different kind of wastes, but the use as well as practice of handling different kind of wastes in accordance to the rules is still inadequate. Waste prevention needs to play a larger role in the local waste management in order to accelerate the transition of wastes and resources management towards a more integrated management (Zacho and Mosgaard, 2016). Pallavi et al. (2017) have recommended the internet of things (IoT) to manage the solid wastes effectively. Waste production in India is huge and the traditional management of the mixed and contaminated wastes is a serious public health issue creating substantial environmental concerns. In order to alleviate the spread of serious contagious diseases, it is required to enforce adequate hospital or biomedical waste management policy, support in strong public health promotional research, media-campaigns, safe-practices, and hygiene education by means of social and mass media (Iyengar and Islam, 2017). Sharma and Gupta (2017) have explored the healthcare waste management circumstances in Himachal Pradesh (India). The private hospitals were found to generate more healthcare wastes in comparison to public hospitals. Aryampa et al. (2019) have revealed the major drivers of waste generations, collections, and disposal in the East African Community as population-growths, vehicle-capacity, and disposal requirements, respectively. The rate of waste generation in Kampala enlarged from 0.26 to 0.47 kg per capita per day, and the annual waste-quantity augmented significantly (i.e., $p< 0.5$) from 227,916 to 481,081 tons (by 48%) corresponding to a 54% increase in population. The solid waste management has been a key concern in the metropolitan cities of the developed as well as developing countries. With the growth in population as the garbage is also increasing, and the huge un-managed accumulation of this garbage pollutes the environment, spoils the beauty of the region and also lead to the health hazards. Waste is one of the present extraordinary ecological difficulties. Much exertion is committed to gather and recover the materials in waste, yet from an ecological point of view, anticipation of waste is desirable over any sort of waste treatment (landfill, vitality recuperation, and reusing). This is on the grounds that manufacturing and treatment of the items and substances that become waste are maintained by a strategic distance. Waste anticipation is characterized as measures taken before a substance, material, or item gets throw away. It includes severe evasion of waste, for example, a decrease of waste generations. The definition additionally incorporates the expansion of the life length of items and reuse, just as the subjective part of decreasing the danger of waste. Prevention of waste can happen in all life-cycle-periods of an item.

Therefore, the administrative structure governing waste prevention covers a considerable rundown of various directives and guidelines. The waste generation control flowchart is as shown in Figure 2.4.

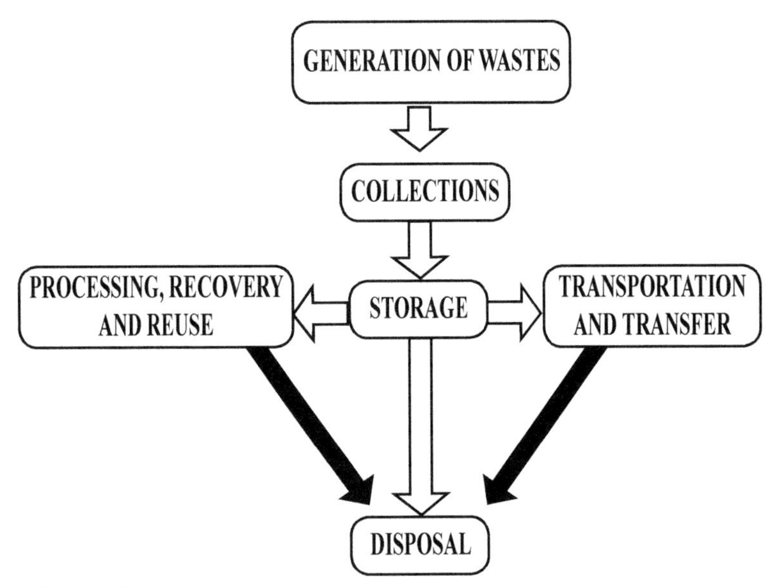

FIGURE 2.4 Waste control.

2.3 WASTE MANAGEMENT SYSTEM

Uniformity in waste management methods cannot be achieved across different regions and sectors, as a singular waste management technique becomes unable to manage all potential waste materials in a reasonable way (Staniškis, 2005). In view of the varying economic, environmental as well as social conditions, there must be flexibility in the waste management systems (McDougall et al., 2001; Scharfe, 2010). Normally, a number of processes are involved in the waste management which are closely interrelated, thus designing holistic waste management systems are logically required instead of alternatives and competing choices (Staniškis, 2005). Waste management frameworks offer flexibility in outlining and investigating quantitative as well as subjective data across various scales, structures to distinguish key objectives and qualities, logic for considering the potential probabilities as well as consequences in relation to some specific option, and communicability to unmistakably convey key plans to key-stakeholders (Owen, 2003). Moreover, an integrated waste management has been developed as an enriched way to deal with managing of wastes by consolidating and applying a scope of reasonable methods, innovations, and management programs to accomplish the explicit-destinations and objectives (McDougall et al., 2001; Tchobanoglous and Kreith, 2002). The idea of integrated waste management was emerged out of the acknowledgment that waste management frameworks are comprised of a number of interconnected frameworks and functions, and has emerged as "structure of planning and executing reference for new waste management frameworks and for dissecting and improving existing frameworks" (UNEP, 1995). Similarly, as there is no individual waste management strategy which is appropriate for preparing all loss in a maintainable way, there is no ideal integrated waste management framework (McDougall et al., 2001). Different classification of wastes based on their sources are as shown in Table 2.1.

TABLE 2.1 Classification of Wastes Based on its Sources

Waste Sources	Basis of Waste Generations	Waste Types
Commercial	Markets, restaurants, hotels, auto-repair shops, motels, printing shops, etc.	Plastic, wood, food-wastes, metallic wastes, glasses, etc.
Residential	Single and multi-family dwellings, apartments of lower, medium, and higher densities, etc.	Wastes related to food, clothing, glasses, card-boards, wood, paper, plastic, consumer electronic products, etc.
Agricultural	Farms, diaries, field crops, orchards, etc.	Agricultural wastes, spoiled foods, hazardous wastes, etc.
Institutional	Government sectors, universities, hospitals, colleges, schools, etc.	All commercial wastes including bio-medical wastes.
Industrial	Power plants, light as well as heavy manufacturing concerns, fabrication works, refineries, chemical industries, etc.	All commercial wastes, including wastes due to industrial processes, and scrap materials.
Construction and Demolition	Newer construction, repairs, and renovations, etc.	Rocks, wood, soils, concretes, steels, gypsum-board, asphalt paving as well as roofing.

Source: Tchobanoglous and Kreith (2002).

The organic as well as inorganic matters are usually contained in MSWs. By adopting suitable waste processing as well as treatment technologies, the recovery of the latent energy of its' organic fractions can be achieved for gainful-utilization. The process in association with different "MSW" components is as shown in Table 2.2.

TABLE 2.2 The Process in Association with Different "Municipal Solid Waste (MSW)" Components

Type of MSW	Components of Wastes	Treatment Process
Biodegradable	Food, garden, and kitchen wastes	Biological-treatments: • Aerobic processes; • Anaerobic processes. Thermal-treatments: • Gasification systems; • Incinerations; • Pyrolysis systems. Transformations: • Mechanical-transformation; • Thermal-transformation.
Recyclable	Papers	• Bleaching process; • De-inking; • Dissolution; • Screening; • Sterilization.
	Glasses	Vitrification technology
	Plastics	• Plasma-pyrolysis-technology (PPT); • Alternate fuel as "refuse-derived-fuel (RDF)."
Inert	Sand	Landfilling: Jaw and pulse crushers
	Pebbles and gravels	

Different wastes constructions and demolitions are as shown in Table 2.3.

The different waste management initiatives taken in India are illustrated in Table 2.4.

The merits and demerits of waste management are as illustrated in Table 2.5.

TABLE 2.3 Wastes from Constructions and Demolitions

Waste Types	Sources of Generation	Components	Treatment Process
Construction-wastes	Wastes from construction and/or maintenance activities of buildings as well as civil infrastructure works.	• Concrete (pre-stressed or normal) cement as well as various mortars conglomerates, tiles, bricks, and blocks. • Excavation of soils, woods, papers, celluloses, and polystyrenes.	Crushing and reuse: Onsite or offsite.
Demolition-wastes	Wastes from maintenance and/or partial or total demolitions of buildings as well as civil infrastructure works.		• Magnetic- separation; • Recycling: On-site or off-site processing for the recovery of high-valuable, saleable-products. • Screen- technology.
Excavation of soils and rocks	Wastes from earth-works for the construction of civil works and/or excavations.	Excavation of soils, woods.	Magnetic-separation, Incineration

TABLE 2.4 Waste Management Initiatives in India

Initiatives	Concerned Issues	
Policies and regulations	Policy initiatives	• Ecomark Scheme (1991); • Law Commission Recommendation; • National Conservation Strategy and Policy Statement on Environment and Development (1992); • Policy Statement for Abatement of Pollution (1992); • National Environment Policy (2006); • National Urban Sanitation Policy (2008).
	Legal frameworks	• Management and handling rules; • 74th Constitutional Amendment Act (1992); • Environment (Protection) Act (1986); • Water (Prevention and Control of Pollution) Act (1974); • Water (Prevention and Control of Pollution) Cess Act (1977); • National Environment Tribunal Act (1995); • National Environment Appellate Authority Act (1997).
	Institutional frameworks	• State levels; • Central levels; • Organizational levels.
	Environmental norms	• Recent notified environmental standards; • Existing environmental standards.
Governmental programs	Campaign on total sanitation	• Scope and structure of program; • Funding; • Experience to date; • Challenges and issues.

TABLE 2.4 *(Continued)*

Initiatives	Concerned Issues	
	JNNURM	• Scope and structure of program; • Funding; • Experience to date; • Challenges and issues; • Expertise on reforms.
	Waste to energy programs by MNREs'	• Scope and structure of program; • Experience to date; • Challenges and issues.
	Other programs	• Integrated lower sanitation costs schemes; • Program on national biogas and manure management.
Practices and technologies	Established technologies	• Sanitations; • Landfills; • Waste incineration.
	Major projects	• Solid waste management improvement project in Kolkata; • Solid waste management in Kanchrapara through citizens' participation; • Municipal solid waste management project in Kollam; • Municipal solid waste management project in Chennai; • Municipal solid waste management project in Navi-Mumbai; • Ultra-modern waste management plant in Gurgaon; • Zero garbage status: Namakkal; • Suryapet: Zero garbage and dustbin free town; • Solid waste management in Visakhapatnam through citizens' participation;

TABLE 2.4 *(Continued)*

Initiatives	Concerned Issues	
	Major initiatives	• Decentralized solid waste management in Thiruvananthapuram; • CIDCO: Solid waste management system at areas adjoining Navi-Mumbai. • GPRS enabled waste-bin in Chennai; • Tapping methane gas in Ahmadabad; • Solid waste management corporation in Goa; • Bye-laws for collecting generated wastes in hotels in Nagpur; • Door to door collection of solid wastes in Yavatmal.
Rural wastes management	Major projects	• Zero-waste management at Vellore district in Tamil Nadu; • Solid waste management at Dhamner village Maharashtra; • Grey-water management at Fathepura village in Gujarat; • Waste-paper to pep-wood in Nasik.

Source: India Infrastructure Report (2009).

TABLE 2.5 Merits and De-Merits of Waste Management

Waste Management Options	Merits	De-Merits
Landfills	• Cheaper disposal technique; • Before reclamation the wastes are used to back-fill quarries; • Contribution of landfill gases to renewable energy supply.	• Leachate and run-off causes water pollution; • Anaerobic decomposition of organic matters to generate carbon dioxide, methane, sulfur, nitrogen, and volatile organic compounds cause air pollution; • Known or suspected emission of carcinogens or teratogens occurs such as polycyclic-aromatic-hydrocarbons, dioxins, vinyl-chloride, benzenes, chromium, nickel, arsenic, etc.; • Causes animal vectors.
Sewage treatments	• Human wastes disposal in safer way.	• Waste discharge may include endocrine-disrupting compounds, organic compounds, pathogenic micro-organisms, heavy metals.
Incinerations	• volume and weight of waste gets reduced, and about 30% is left as ash that can be utilized for recovery of materials; • Sources of portable water supplies are protected; • Helps to generate electricity; • Helps to reduce the potential infectivity of clinical wastes.	• Hazardous solid wastes are produced; • Nuisance of odors are produced; • Heavy metals, toxic pollutants, and products of combustion are emitted; • Contaminated wastewater gets discharged.
Composting	• Reduction of wastes for disposing to landfills and incineration; • Useful organic matters are recovered for use as soil-amendments; • Generates opportunities of employments.	• Nuisance of odors, noises, and vermin; • Presence of bio-aerosols organic dusts comprising of bacteria or fungal-spores; • Emission of volatile organic compounds; • Creation of potential pathways for contaminants to get into food-chains by the utilization on lands.

TABLE 2.5 *(Continued)*

Waste Management Options	Merits	De-Merits
Re-cycling	• Helps in resources conservation; • Helps to supply raw-materials for industrial purposes; • Reduction of wastes for disposing to landfills and incineration.	• Existence of a variety of processes; • Harmful emissions; • Utilization of more energy for processes; • Lower demand for product at present; • Requirement of co-operation of individuals.

Source: Rushton (2003).

An extra consideration beyond the ecological impacts has been a huge requirement in order to plan the waste disposal, treatments, services, and facilities. For the sustainable use of the facilities in waste management, the decision-making requires a tradeoff between a number of conflicting objectives, as growing of one benefit could result in decreasing the others. Any waste management framework can be considered as an optimization process if the socio-cultural, environmental, political, economic, and technical dimensions are taken into account that reduces the conflict among stakeholders and integrates this reality. A number of studies has been carried out by using the multi-criteria decision method/analysis for resolving the waste management issues (Abba et al., 2013; Debnath et al., 2017; Hanan et al., 2013; Karagiannidis et al., 2010; Korucu and Erdagi, 2012; Si et al., 2018). Through proper and effective waste management actions, not only the existing waste management systems will be complemented, but also a comprehensive and integrated waste management plan for the long term can be developed for future as well as further benefits. Based on the strength, weakness, opportunities as well as threats (SWOT) analysis of the existing technologies (Table 2.6) in waste management initiative in the Indian context, it is clear that there is still a requirement for an effective waste management framework to overcome the possible weaknesses and threats.

The waste management practices must include the steps as summarized in Figure 2.5.

2.4 CONCLUSION

Many factors like growth in population, competitiveness in economy, rise in living standards of communities, and urbanization have resulted in an exponential growth rate in the wastes in developing countries. If the wastes are not properly managed, then it can cause adverse impacts on the environment, community-healthiness, and other socio-economic factors. A long-term deficient disposal of MSW will cause poor governance and standard of living, poverty, major health-problems, social, and environmental issues, are resulted because of long-term and deficient waste management systems. Therefore, the selection of appropriate wastes disposal and treatment methods have been a complex decision-making problem that needs to be focused on from all levels. Generally, these kinds of decision-making problems are poorly-structured, and relatively the researchers have un-successfully utilized their human-intelligence to solve such issues. Moreover, a

TABLE 2.6 SWOT Analysis of Existing Technologies in Waste Treatment

Waste Treatment Methods	Strength	Weakness	Opportunities	Threats
Dry-composting	• Dry-wastes can be preserved for use in future; • The bio-logical process to take place in an open or confined area; • Possibilities of obtaining nutrients rich organic fertilizers as well as soil conditioners from wastes.	• Capability of managing only bio-degradable wastes; • Difficulty in\emission control.	• Options of resource recovery and to make bio-fertilizers; • Options of generating biogas from dry-wastes.	• Water and soil contaminations due to poor management; • Environmental impacts of emission.
Sanitary-landfills	• Management of waste in controlled environment; • A natural decomposition process capable of handling various kinds of wastes with larger volumes.	• Requirement of huge land areas; • Longer time for reclaiming the landfill lands restorations.	• Options of recovering biogas from landfills; • More environment friendly management of wastes.	• Air, water, and soil contaminations due to poor management.
Plasma-arc	• Most of the wastes can be handled with lower disposable residues.	• Higher investment costs.	• Options of recovering higher energy and heat.	• Environmental impacts of emission.
Pyrolysis	• Most of the wastes can be handled with lower disposable residues.	• Higher investment costs.	• Options of recovering resources and energy.	• Environmental impacts of emission.
Gasification	• Most of the wastes can be handled with lower disposable residues.	• Higher investment costs.	• Options of recovering energy and heat.	• Environmental impacts of emission.

transparent, systematic, and documented decision-making process can make an effective framework in appropriate waste management in India.

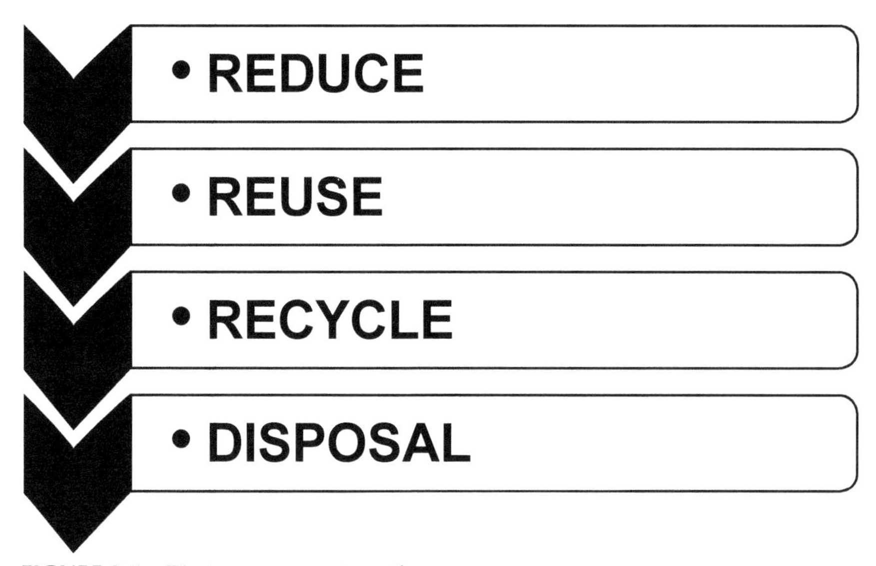

FIGURE 2.5 Waste management practice.

KEYWORDS

- **domestic wastes**
- **integrated waste management**
- **SWOT analysis**
- **urbanization**
- **waste management**
- **waste treatment**

REFERENCES

Abba, A. H., Noor, Z. Z., Yusuf, R. O., Din, M. F. M. D., & Hassan, M. A. A., (2013). Assessing environmental impacts of municipal solid waste of Johor by analytical hierarchy process. *Resour. Conserv. Recycle., 73,* 188–196.

Abdulla, F., Qdais, H. A., & Rabi, A., (2008). Site investigation on medical waste management practices in northern Jordan. *Journal of Waste Management, 28*(2), 450–458.

Alves, L., Freire, L., & Vazquez, E., (2016). Waste management in major events and the relevance of environmental certification. *Proceedings of the 8 International Conference on Waste Management and the Environment (WM 2016), WIT Transactions on Ecology and the Environment, 202,* 49–56. doi: 10.2495/WM160051.

Annepu, R. K., (2012). *Report on Sustainable Solid Waste Management in India* (pp. 1–189). Waste-to-Energy Research and Technology Council (WTERT). Available at: http://swmindia.blogspot.in/ (accessed on 18 February 2021).

Aryampa, S., Maheshwari, B., Sabiiti, E., Bateganya, N. L., & Bukenya, B., (2019). Status of waste management in the East African cities: Understanding the drivers of waste generation, collection and disposal and their impacts on Kampala city's sustainability. *Sustainability, 11,* 5523. doi: 10.3390/su11195523.

Bogner, J., et al., (2007). Waste management. In, Metz, B., et al., (eds.), *Contribution of Working Group III to the Fourth Assessment Report of the Intergovernmental Panel on Climate Change* (pp. 586–618). Chapter 10. Cambridge University Press, Cambridge, United Kingdom and New York, NY, USA.

Census, (2011). *Provisional Population Totals.* India. Retrieved from: https://censusindia.gov.in/2011-prov-results/data_files/india/pov_popu_total_presentation_2011.pdf (accessed on 18 February 2021).

Central Pollution Control Board of India (CPCB), (1998). *Status of Solid Waste Management in Metro Cities.* CPCB.

Chancerel, P., Meskers, C. E., Hagelüken, C., & Potter, V. S., (2009). Assessment of precious metal flows during preprocessing of waste electrical and electronic equipment. *Journal of Industrial Ecology, 13*(5), 791–810.

Debnath, A., Roy, J., Kar, S., Zavadskas, E. K., & Antucheviciene, J., (2017). A hybrid MCDM approach for strategic project portfolio selection of agro by-products. *Sustainability, 9,* 1302. doi: 10.3390/su9081302.

Dinesh, M. S., Geetha, K. S., Vaishnavi, V., Kale, R. D., & Krishna, M. V., (2010). Ecofriendly treatment of biomedical wastes using epigenic earthworms. *Journal of ISHWM, 9*(1), 5–20.

Dwivedi, A. K., Pandey, S., & Shashi, (2009). Fate of hospital waste in India. *Biology and Medicine, 1*(3), 25–32.

European Environment Agency, (2007). *The Road from Landfilling to Recycling: Common Destination, Different Routes.* Brochure No 4/2007., Denmark. Available at: http://www.eea.europa.eu/publications/brochure_2007_4 (accessed on 18 February 2021).

Federation of Indian Chambers of Commerce and Industry, (2009). *Survey on the Current Status of Municipal Solid Waste Management in Indian Cities and the Potential of Landfill Gas to Energy Projects in India.* http://www.hpccc.gov.in/PDF/Solid_Waste/Current%20Status%20of%20MSW.pdf (accessed on 18 February 2021).

Giusti, L., (2009). A review of waste management practices and their impact on human health. *Waste Management, 29,* 2227–2239.

Hanan, D., Burnley, S., & Cooke, D., (2013). A multi-criteria decision analysis assessment of waste paper management options. *Waste Manag., 33,* 566–573.

Huisman, J., Magalini, F., Kuehr, R., Maurer, C., Ogilvie, S., Poll, J., Delgado, C., Artim, E., & Szlezak, A., (2007). *Review of Directive 2002/96 on Waste Electrical and Electronic Equipment (WEEE)* (p. 377). Bonn: United Nations University.

ITU, (2011a). *Measuring the Information Society*. A report from the International Telecommunication Union. https://www.itu.int/net/pressoffice/backgrounders/general/pdf/5.pdf (accessed on 18 February 2021).

ITU, (2011b). The world in 2011. *ICT Facts and Figures*. http://www.itu.int/ITU-D/ict/facts/2011/material/ICTFactsFigures2011.pdf (accessed on 18 February 2021).

Iyengar, V., & Islam, M. R., (2017). Biomedical, sharps and general waste disposal in India: Potential for the spread of contagious diseases and serious environmental contamination. *Universal Journal of Public Health, 5*(5), 271–274. doi: 10.13189/ujph.2017.050509.

Joshi, R., & Ahmed, S., (2016). Status and challenges of municipal solid waste management in India: A review. *Cogent Environmental Science, 2,* 1139434.

Karagiannidis, A., Papageorgiou, A., Perkoulidis, G., Sanida, G., & Samaras, P., (2010). A multi-criteria assessment of scenarios on thermal processing of infectious hospital wastes: A case study for central Macedonia. *Waste Manag., 30,* 251–262.

Korucu, M. K., & Erdagi, B., (2012). A criticism of applications with multi-criteria decision analysis that are used for the site selection for the disposal of municipal solid wastes. *Waste Manag., 32,* 2315–2323.

Lee, G. F., & Jones-Lee, A., (1994). *Impact of Municipal and Industrial Non-Hazardous Waste Landfills on Public Health and the Environment: An Overview* (pp. 15–18). G. Fred Lee & Associates: California, CA, USA.

Liyala, C. M., (2011). *Modernizing Solid Waste Management at Municipal Level: Institutional Arrangements in Urban Centers of East Africa*. PhD Thesis. Environmental Policy Series. Wageningen University. The Netherlands.

Manyele, S. V., & Lyasenga, T. J., (2010). Factors affecting medical waste management in low level health facilities in Tanzania. *African Journal of Environmental Science and Technology, 4*(5), 304–318.

Manzurul, H. M., Ahmed, S. A., Rahman, A. K., & Biswas, T. K., (2008). Pattern of medical waste management: existing scenario in Dhaka City, Bangladesh. *Journal of BMC Public Health, 8,* 36.

Mbeng, L. O., Phillips, P. S., & Fairweather, R., (2009). Developing sustainable waste management practice: Application of Q methodology to construct new strategy component in Limbe-Cameroon. *The Open Waste Management Journal, 2,* 27–36.

McDougall, F. R., White, P. R., Franke, M., & Hindle, P., (2001). *Integrated Solid Waste Management: A Life-Cycle Inventory* (2nd edn., p. 544). Blackwell.

Oberlin, A. S., (2011). *The Role of Households in Solid Waste Management in East Africa Capital Cities.* PhD Thesis. Environmental Policy Series. Wageningen University. The Netherlands.

Okot-Okumu, J., & Nyenje, R., (2011). Municipal solid waste management under decentralization in Uganda. *Habitat International, 35,* 537–543.

Owen, R., (2003). *Preparing a Recommendation to Governments on Cleanup Options for the Sydney Tar Ponds and Coke Ovens sites: An Evaluation of Environmental Decision-Making Tools.*

Pallavi, K. N., Ravi, K. V., & Chaithra, B. M., (2017). Smart waste management using internet of things: A survey. *International Conference on I-SMAC (IoT in Social, Mobile, Analytics and Cloud) (I-SMAC 2017)*, 60–64.

Patil, A. D., & Shekdar, A. V., (2001). Healthcare waste management in India. *Journal of Environmental Management, 63,* 211–220.

Planning Commission, Government of India, (2014). *Report of the Task Force on Waste to Energy in the Context of Integrated Municipal Solid Waste Management* (Vol. I). Available at: http://swachhbharaturban.gov.in/writereaddata/Task_force_report_on_WTE.pdf (accessed on 18 February 2021).

Pope, J., Annandale, D., & Morrison-Saunders, A., (2004). Conceptualizing sustainability assessment. *Environ. Impact Assess. Rev., 24,* 595–616.

Rana, R., Ganguly, R., & Gupta, A. K., (2015). An assessment of solid waste management system in Chandigarh city India. *Electron. J. Geotech. Eng., 20,* 1547–1572.

Rushton, L., (2003). Health hazards and waste management. *British Medical Bulletin, 68,* 183–197. doi: 10.1093/bmb/ldg034.

Scharfe, D., (2010). *Integrated Waste Management Plan.* Center & South Hastings waste services board/waste diversion Ontario and Stewardship Ontario.

Scheinberg, A., (2011). *Value Added: Modes of Sustainable Recycling in the Modernization of Waste Management Systems.* PhD Thesis. Wageningen University. The Netherlands.

Schluep, M., Hagelueken, C., Kuehr, R., Magalini, F., Maurer, C., Meskers, C., Mueller, E., & Wang, F., (2009). *Sustainable Innovation and Technology Transfer Industrial Sector Studies* (p. 120). Recycling from e-waste to resources. United Nations Environment Program & United Nations University.

Sharma, S. K., & Gupta, S., (2017). Healthcare waste management scenario: A case of Himachal Pradesh (India). *Clinical Epidemiology and Global Health, 5,* 169–172.

Si, S. L., You, X. Y., Liu, H. C., & Zhang, P., (2018). DEMATEL technique: A systematic review of the state-of-the-art literature on methodologies and applications. *Mathematical Problems in Engineering,* Article ID: 3696457. From https://doi.org/10.1155/2018/3696457.

Staniškis, J., (2005). Integrated waste management: Concept and implementation. *Environmental Research, Engineering and Management, 3*(33), 40–46.

Tchobanoglous, G., & Kreith, F., (2002). *Handbook of Solid Waste Management* (2nd edn., p. 950). McGraw-Hill.

The United Nations for Environmental Program, (2010). *Waste and Climate Change: Global Trends and Strategy Framework.* UNEP, Division of Technology Industry and Economics, International Environmental Technology Center, Osaka/Shiga.

U.S Environmental Protection Agency, (1995). *Decision-Makers' Guide to Solid Waste Management* (Vol. II). Washington, D.C. Retrieved from: http://www.epa.gov/osw/nonhaz/municipal/dmg2/ (accessed on 18 February 2021).

UNEP News Release, (1999). *Ministers Call for Cleaner Production Methods as They Set Priorities for Next Decade of Basel Convention on Hazardous Wastes.* http://www.unep.org/ (accessed on 18 February 2021).

Verma, L. K., (2010). Managing hospital waste is difficult: How difficult? *Journal of ISHWM, 9*(1), 46–50.

Widmer, R., Oswald-Krapf, H., Sinha-Khetriwal, D., Böni, H., & Schnellmann, M., (2005). Global perspectives on e-waste. *Environmental Impact Assessment Review, 25*(5), 436–458.

Wilson, D. C., Velis, C., & Cheeseman, C., (2006). Role of informal sector recycling in waste management in developing countries. *Habitat Int., 30,* 797–808. doi: 10.1016/j.habitatint.2005.09.005.

Zacho, K. O., & Mosgaard, M. A., (2016). Understanding the role of waste prevention in local waste management: A literature review. *Waste Management and Research,* 1–15. doi: 10.1177/0734242X16652958.

Zaman, A. U., (2013). Identification of waste management development drivers and potential emerging waste treatment technologies. *Int. J. Environ. Sci. Technol., 10,* 455–464. doi: 10.1007/s13762-013-0187-2.

Zheng, S., Liu, W., & Zhi, Q., (2016). Cleaner waste management: A review based on the aspects of technology, market and policy. CUE2016-applied energy symposium and forum 2016: Low carbon cities and urban energy systems. *Energy Procedia, 104,* 492–497.

CHAPTER 3

PRIORITIZATION OF BARRIERS OF WASTE MANAGEMENT: A PUBLIC-AWARENESS APPROACH

ABSTRACT

In the present situation, waste is contaminating the world, which may be a significant concern to take care of a safe and clean environment. However, this issue is often addressed by transfer and bona fide treatment of wastes, both irresistible and infectious. So, the message of an unpolluted environment is increasing everywhere on this planet. For this reason, efforts are made in this research to get past the boundaries that limit the most uncomplicated transfer, so the best treatment of wastes can be achieved in India. A broad examination is completed on various sorts of wastes like hospital or bio-medical wastes (HBW), municipal solid wastes (MSW), paper-wastes (PW), and e-wastes (EW), followed by the finding of some basic obstructions for every waste. Subsequently, all hindrances are positioned in consistent with their importance by "VIKOR analysis" to such an extent that the essential moves can be taken care of for India's waste management.

3.1 INTRODUCTION

Waste are some things which vary counting on the geographic-locations and areas. These wastes need to be disposed of and treated properly, in order to protect the world and keep it healthy. Various methods used for the disposal of waste include: landfill, combustion or incineration, composting, plasma-gasification, composting of toilets, aerated static-pile composting, anaerobic digestion (AD), bio-drying, pyrolysis, mechanical-biological treatment-systems and windrow-composting, respectively. Apart from of these, there are "3R's: Reduce, Reuse, and Recycle," which assumes an important-role with the management of wastes. In-consistent with the "metropolitan solid waste (management and handling) rules (2000)," the duty of municipal corporation includes "collection, isolation, transportation, processing, and arrangement of the municipal solid wastes (MSWs) in an environment-friendly as well as scientific manner. The proper treatment of wastes provides subsequent benefits like exclusion of health-hazards, reducing the breeding of flies, rodents, and mosquitoes, peoples' lifestyle betterment, taking away of foul-odors, contamination abolition of any soil or water, renewable-energy benefits, and natural-resources conservation. Wastes management take care of all the performances and behaviors necessary for the management of wastes from its' sources to final-disposal. Several Government-policies along with a number of researches have already been performed on the management of wastes. But still, the difficulties of wastes management along with proper disposals issues has remained unresolved. So an attempt was taken in this study to look for the common-barriers in the management for various wastes' categories.

The different types of generated wastes in India include the following:

1. **Paper-wastes (PW):** Paper is produced by the utilization of cellulosic fibers which are acquired from various plants. After the use of papers, these are tossed into junk. Paper-industry are probably the foremost seasoned industries that have caught the Indian-market. There has been a developing interest in paper and paper-board which may be a quantum jump. The crude materials for paper industry vigorously depend upon imported papers. Within the event that the paper assortment component and reusing of paper are expanded, there will be lesser import of papers, decrease in ecological-issues and business-age. Numerous NGO's have entered within the assortment and reusing program for thrown away or waste papers. An office assortment instrument needs to be actualized to collect papers from work-places, and many NGO's have started this program. Shredder needs to be suggested for personal papers within the work-places and ventures. Projects need to be led in schools for collection of note-books and recycling processes. Additionally, the teachers should be involved in teaching their students about the reduction, reuse, and recycling

of papers. However, an outsized portion of the state has begun reusing of paper and paper assortment system, which are worked and composed by city specialists who are bolstered by their residents. In this way, the assortment-system of paper is exceptionally fruitful. Some countries also send an enormous measure of papers after they meet their household pre-requisites. As per "Indian Railways Catering and Tourism Corporation," around 3 lakhs A4-size papers are often spared regularly by the use of tickets' electronic-version. For the generation of 1 ton of paper, 17 trees and 7000 gallons of water are required. Additionally, 1 tree gives oxygen to very nearly 3 people. The PW collection mechanisms in India and the recovery potential for PW are shown in Tables 3.1 and 3.2, respectively.

TABLE 3.1 Present Paper-Wastes Collection Mechanisms in India

Sources	Collected-Items	Collected-by	Quantity Collected (in Million-Tones Per Annum)
Collection from households	Old newspapers and magazines, notebooks, and textbooks	Weekend-Hawkers	2.00
Annual scrap-contracts of publishers, printers, and converters	Paper-trimmings, print-rejects, misprint or overprint sheets and other wastes	Contractors	0.25
Scrap-contracts with offices, industries, and libraries	Old corrugated-cartons, examination answer-sheets, library-records, old-office documents	Contractors	0.50

Source: ITC-WOW (2011).

2. **Municipal solid wastes (MSW):** The disposal of solid wastes on land results in the emission of harmful greenhouse gases, and increased breeding rate of pests and rodents, the inflammable gases generation like methane under the waste-dumps, and an added soil-acidity. Moreover, the leachates produced by the wastes will contaminate the ground-water. The solid waste management institutions and their functions are illustrated in Table 3.3.

TABLE 3.2 Recovery Potential for Paper-Wastes

Paper-Grades (Writing or Printing)	Potential-Sources of Generation	Generation or Consumption (%)	Type of Waste	Collection Rate (%)
Copier-papers	• Offices	50	Post-consumer	20
	• Business-establishments	40		
	• Others	10		
Cream-wove	• Printing-house	20	Pre-consumer	100
	• Paper-traders	5		
	• Households	20	Post-consumer	20
	• Schools or Colleges	10		
	• Offices	25		
	• Business-establishments	10		
	• Others	10		
Packaging-papers	• Converting-houses	15	Pre-consumer	100
	• Households	20	Post-consumer	50
	• Offices	5		
	• Business-establishments	50		
	• Others	10		
News-papers	• Publishing-houses	20	Pre-consumer	100
	• Distributors	5		
	• Households	40	Post-consumer	30
	• Offices	10		
	• Business-establishments	15		
	• Others	10		

Source: Indian Recycled Paper Mills Association (IRPMA) (2011).

TABLE 3.3 Solid Waste Management Institutions and Functions

Responsible-Institution	Roles and Responsibilities in Solid Waste Management
Central-Government	Formulating laws and rules, framing of policies, preparing of guidelines, manuals, and technical assistance, providing financial-supports, monitoring of implemented laws and rules
State-Government	Formulating state-level laws and rules, framing of policies, preparing of guidelines, manuals, and technical assistance, providing financial-supports, monitoring of implemented laws and rules
Municipal-Authorities and State-Government	Planning for SWM treatment-facilities
Municipal-Authorities	Collection, transportation, treatment, and disposal of wastes
Municipal-Authorities with the approval of State-Government	Framing by-laws, levy, and collection of fees
Municipal-Authorities and State and Central Governments	Financing of SWM systems

Source: Improving municipal solid waste management in India: A sourcebook for policymakers and practitioners, World Bank Institute.

3. **Hospital and biomedical wastes (HBW):** Emergency-clinics are the places which are, for the most part, visited by wiped out individuals. Medical-clinics produce an enormous amount of perilous and irresistible waste which if not taken consideration will in the long run become a risk towards open. In the current situation, as a result of populace blast, there is a gigantic age of bio-medical wastes. Appropriate treatment of these wastes ought to be our objective. Bio-medical wastes are equipped for transmitting transferable infections. In this way, appropriate administration and treatment of these wastes have turned out to be an essential requirement. The different categories of hospital/bio-medical wastes and their disposal methods are as illustrated in Table 3.4. Similarly, the color-coding and types of containers for disposals of bio-medical wastes are as illustrated in Table 3.5.
4. **E-Wastes (EW):** These days, the use of electronic and electrical products particularly have radically expanded for the smoother, easiest, and economical life. In any case, this expansion in innovation and utilization of electronic products is prompting a generation

TABLE 3.4 Categories of Bio-Medical/Hospital Wastes and Their Disposal Methods (Schedule-I)

Waste-Category Number	Waste-Category Type	Treatment and Disposal
1.	Human-anatomical wastes: human-tissues, organs, body-parts	Incineration@/deep-burial
2.	Animal-wastes: animal-tissues, organs, body-parts carcasses, bleeding-parts	Incineration/deep-burial
3.	Microbiology and biotechnology wastes: wastes from laboratory-cultures, stocks or specific specimen of micro-organisms, live or attenuated vaccines, animal, and human cell-culture used in research, infectious agents from research and industrial laboratories, and wastes from production of biological, toxin, dish, and device used for culture-transfer	Local-autoclaving/microwaving/ incineration
4.	Waste-sharps: blades, needles, scalpels, syringes, glass, etc., that may cause sharp punctures and cuts, and includes both unused and used sharps.	Disinfection (chemical-treatment/ microwaving/autoclaving and mutilation/shredding
5.	Discarded-medicines and cytotoxic-drugs: Wastes comprising of outdated, contaminated, and discarded medicines	Incineration/destruction and drugs-disposal in secured-landfills
6.	Soiled wastes: contaminated-items with bloods and body fluids including cottons, dressings, soiled plaster-casts, linens, beddings, other materials contaminated with bloods	Incineration/microwaving/ autoclaving
7.	Solid wastes: wastes generated from disposable-items other than waste-sharps such as catheters, tubing, intravenous-sets, etc.	Disinfection by chemical-treatment/ microwaving/autoclaving and mutilation/shredding
8.	Liquid wastes: wastes generated from laboratories and washings, cleanings, house-keeping, and disinfecting activities	Disinfection by chemical-treatment and discharges into drains
9.	Incineration-ashes: ashes from any bio-medical wastes' incineration	Disposals in municipal-landfill
10.	Chemical-wastes: chemical utilized in production of biological, disinfection, insecticides, etc.	Chemical-treatment and discharges into drains for liquids and secured-landfill for solids

Source: Central Pollution Control Board, India.

of an incredible mass of EW also. Moreover, there is just halfway recyclability of the hardware which is tossed-out due to its' expiry. In this way, it is a critical perspective to affect the management of EW. There are a variety of EW from urban or industrial wastes that contain both dangerous and valuable materials requiring special-treatment and recycling-processes in order to avoid the adverse environmental impacts, and harmful-impacts on human-health. Recovering of base-metals is conceivable by reusing e-squander, yet the highest labor-costs and therefore, the exacting ecological-enactment have united these exercises' usage for the foremost part in Asian countries like China and India by utilization of out-of-date techniques and lacking accentuation on the employees' protection. Due to this, the EW disposal-issues has attracted the interest of politicians, and various non-governmental organizations.

TABLE 3.5 Color-Coding and Types of Container for Disposals of Bio-Medical Wastes (Schedule-II)

Color-Coding	Types of Container	Waste-Category Number	Treatment Options as Per Schedule-I
Yellow	Plastic-bags	1, 2, 3, and 6	Incineration/deep-burial
Red	Disinfected-containers/ plastic-bags	3, 6, and 7	Autoclaving/microwaving/ chemical-treatment
Blue/white translucent	Plastic-bags/puncture-proof containers	4 and 7	Autoclaving/microwaving/ chemical-treatment and destruction/shredding
Black	Plastic-bag	5, 9, and 10 (solid)	Disposals in secured-landfill

Source: Central Pollution Control Board, India.

Notes:

- Color-coding of plastic categories with multiple-treatment options as defined in Schedule-I, shall be selected depending on chosen treatment-options, which shall be specified in Schedule-I.
- Waste-collection bags for waste types that require incineration shall not be made of chlorinated-plastics.
- Non-requirement of containers/bags for categories 8 and 10 (Liquid).
- Non-requirement of containers/bags for category 3, if disinfected locally.

A number of materials are discarded as wastes after its use. The abandoned tires, discarded waste materials, and paper wastes are shown in Figures 3.1–3.3, respectively.

FIGURE 3.1 Abandoned tires.

FIGURE 3.2 Discarded wastes creating environmental pollution.

FIGURE 3.3 Paper wastes.

The mismanagement of solid wastes has been a global-issue in view of economic-sustainability, environmental-contamination, and social-inclusion (Gupta et al., 2015; Vitorino de Souza Melaré et al., 2017), for which there is need for holistic-approaches and integrated-assessments (Bing et al., 2016). For developing and transition countries attention should be made on un-sustainable management-systems (The World Bank, 2012). There is a need to highlight the differences among developing and rural areas with different management issues, specifically regarding the waste's generation amounts and the available facilities for management of solid wastes (Ferronato et al., 2019). However, negative economic-legislative, technical, political, and operational-limitations occur for both the cases (Imam et al., 2008). In the case of water, plants, and soil, the uncontrolled-disposal generates serious heavy-metals pollution (Vongdala et al., 2019), different pollutant emissions affecting the atmosphere such as "CO, CO_2, SO, NO" are due to open-burning (Wiedinmyer et al., 2014), people working on the areas of waste-picking within open-dump sites are posed to serious health-risks (Gutberlet and Baeder, 2008), the enhancement of environmental-contamination are caused by the release of solid wastes in water-bodies that improve the marine-litter globally (UNEP-Global, 2009). This solid wastes mismanagement is the major reason for various environmental as well as social impacts not allowing any improvement in sustainable-developments. By changing the way of "produce-consume-waste" of products and resources helps in achieving both economic and sustainable-development (Sustainable Development Goals, 2016). Although a significant growth was occurred in the living standards by the growth of the material-footprints of developing-countries from "5 t inh^{-1} in 2000 to 9 t inh^{-1} in 2017," in the national-regulations, its' sustainable management is still not included (UN, 2018). With the introduction of the principles of sustainable development within the "sustainable development goals (SDGs)" having 17 objectives to reduce poverty, improve social-equality, decrease environmental-pollution and to ameliorate city-livability, the global waste management goals in particular for improving sustainability at global-levels are: by 2020, an assurance of the access for all to sufficient, safer, and affordable solid wastes collection services, stoppage of uncontrolled-dumping as well as open-burning, achievement of sustainable and environmental sound management by 2030 for all hazardous-wastes (Wilson and Velis, 2015). Several studies have reported of possible solutions for substantial improvements in the solid waste management in developing countries. For example, buyback-programs for organic-wastes with production of compost or bio-gases (Hettiarachchi, 2018), waste-to-energy technologies and plans implementation (Ouda et al., 2016), in parallel with recycling of glasses, metals, and other inerts through waste-to-energy plans (Sadef et al., 2016), creation of energy from biomass-wastes by briquettes-making (Sawadogo et al., 2018), integrating legal incentives with waste pickers (Ghisolfi et al., 2017), and so on. However, many barriers are still remaining for improving the formal collections, treatments, and final-disposals (Matter et al., 2015). A number of studies have been done in regard to solid waste management in developed as well as developing countries, and concerning the environmental contamination from wastes. For instance, about production of char-fuels (Lohri et al., 2016), "waste electric and electronic equipment (WEEE)" management (Ongondo et al., 2011), "food-waste management" (Thi et al., 2015), and treatments (Lim et al., 2016), used-batteries recycling (Bernardes et al., 2004), informal-sectors inclusion (Ezeah et al., 2013), and the risks posed to vulnerable informal-workers (Brown and McGranahan, 2016), solid waste management associated atmospheric-pollution (Tian et al., 2013), management of household hazardous-wastes (Inglezakis and Moustakas, 2015), and management of health-care wastes (Ali et al., 2017), and so on. Many environmental as well as health-impacts are primarily because of open-dumping and open-burning of wastes that are implemented for waste treatments and disposals in developing-countries (Ferronato et al., 2017; Yukalang et al., 2017; Vaccari et al., 2018). The un-sustainable practices involved in every waste fractions, such as MSWs, health-care wastes, construction, and demolition wastes, used tires and batteries, and industrial wastes, are supposed to spread specific contaminant-concentrations to soil, air, and water. The waste-pickers need to work within these sites in order to collect recyclable-materials that can be sold in local markets, although these informal practices allow in reducing the waste inflow-amounts into open-dumps and water-bodies (Sasaki and Araki, 2014; Linzner and Salhofer, 2014), and the improvement of health and occupational risks are posed by these hazardous activities (Gutberlet et al., 2013; Singh et al., 2018). As most of the peoples are getting affected by the smokes from burning debris as well as by the smell of decomposing of wastes, thus these have become a challenging-issue (Modak et al., 2015), with the concern for un-controlled disposals in open-damps and spaces creating adverse public-health issues (Manaf et al., 2009).

It was reported in a study that the location of Banjul (Gambia) dump-site was in a densely populated area visible to the occupants (Sanneh et al., 2011). Moreover, in "Cambodia" in the capital city "Phnom-Penh," the MSWs were reported of about 3,61,000 tons in 2008, and 6,35,000 tons in 2015 (Seng et al., 2018). In 2004, about 4500 tons of solid wastes were received per day by the landfills of Bangkok in Thailand (Chiemchaisri et al., 2007). The generation of MSWs was reported of about 2728 tons per day, in the West Bank Palestinian territory, in 2005 (Al-Khatib et al., 2015). While, more than 250,000 tons of wastes were generated per year in 'Abuja' city of Nigeria, in 2010 (Aderoju et al., 2018). About 1,200,000 inhabitants with about 0.5 kg of wastes per inhabitants have been reported of generated daily, in 'Maputo' of Mozambique (dos Muchangos et al., 2015). The issues associated with solid waste management has been common globally, with environmental-burdens as well as hazards for the population. The emission of concentration of organic carbons, chlorides, ammonia, heavy-metals take place in the landfill-sites (Torretta et al., 2017), along with higher concentration of chlorides, fluorides, ammonium-nitrogen, "chemical-oxygen-demand (COD)," "biological-oxygen-demand

(BOD)" (Karak et al., 2013). For instance, the MSW dumped at "The Mathkal dump-site" at Kolkata (India) has been reported to be affecting and degrading the water quality in and around dump-site areas (Babu et al., 2010). The soil and water pollution due to open-dumping with the limits imposed by international organizations for soil and water quality was illustrated in Table 3.6.

3.1.1 LITERATURE

An ecological as well as economic solution for waste disposals and treatment should be the minimization of wastes-generation. Although different researches are being done on wastes management, but still there has been a requirement of more attention and research on this concern. By looking at the statistics about wastes across industries (UNEP, 2011, 2013; World Bank, 2012), it was observed of an estimated amount of 1.3 billion-tons of solid-wastes to be collected globally every-year, which is expected to increase to 2.2 billion-tons by 2025 from the developing-countries. About 5% of the global greenhouse-gases are contributed by decomposing of the organic-fractions of solid wastes. Moreover, the global waste-market has been estimated to be US$410 billion a year from collection to recycling, not including the sizable informal-segment in developing-countries. It has been revealed that a ton of aluminum saves 1.3 tons of bauxite-residues, 0.86 m^3 of process-water, 15 m^3 of cooling-water, and 37 oil-barrels, while with the prevention 11 kg of sulfur dioxide and 2 tons of carbon-dioxide emissions. A total of 229,286 jobs were created by recycling activities in the European Union (EU) in 2000, which increased to 512,337 by 2008 with an annual growth rate of 10.57%. Moreover, an increment from 422 persons-per-million inhabitants in 2000 to 611 persons in 2007 was observed with an increase of 45% for the people employed in waste-recovery related activities in Europe. Looking at the food-wastes, about one-third of produced food for human-being consumption is 'wasted or lost' globally, which amount to about 1.3 billion tons per year. About 1 ton of electronic wastes (EW) were found to contain as much gold as 5 to 15 tons of typical gold-ores, and amounts of aluminum, copper, and other rare-metals, which are reported of exceeding by many-times than in typical ores-levels found. The "Printed circuit-boards" are probably the richest ore-streams to ever findings (Grossman, 2006: p. 217). During a study of hospital-wastes at different-levels like national as well as international,

TABLE 3.6 The Concentration of Contaminants in Soil, Run-Off, and Ground-Water Owing to Open-Dumping

Place (Country)	Polluted Environment	Pollutants (Concentrations)	Limits	References
Tiruchirappalli (India)	Soil (mg-kg^{-1})	Pb (291.3>)	50	Kanmani and Gandhimathi (2012)
		Mn (171.16)	500	
		Cd (47.7>)	4	
	Ground-water (mg-L^{-1})	Cu (0.6–2.7)	2	
		Cd (0.16–1.04>)	0.003	
		Mn (0.2–1.8>)	0.4	
		Pb (0.8–5.1>)	0.01	
Havana (Cuba)	Soil (mg-kg^{-1})	Cobalt (8.4)	20	Díaz Rizo et al. (2012)
		Ni (50>)	30	
		Cu (252>)	100	
		Zn (489>)	50	
		Pb (276>)	50	
Nonthaburi (Thailand)	Run-off (mg-L^{-1})	Cr (0.99>)	0.05	Prechthai et al. (2008)
		Cd (0.01>)	0.003	
		Mn (0.49>)	0.4	
		Pb (0.1>)	0.01	
		Ni (0.5>)	0.07	
		Zn (1.32)	4	
		Cu (0.63)	2	
		Hg (0.95>)	0.002	

TABLE 3.6 *(Continued)*

Place (Country)	Polluted Environment	Pollutants (Concentrations)	Limits	References
Uyo (Nigeria)	Soil ($mg\text{-}kg^{-1}$)	Zn (137–146>)	50	Ihedioha et al. (2017)
		Pb (9.9–11.8)	50	
		Ni (11.8–12.6)	30	
		Mn (91.2–94)	500	
		Cr (3.6–4.1>)	1	
		Cd (9.05–12.2>)	4	
Sepang (Malaysia)	Ground-water ($mg\text{-}L^{-1}$)	COD (2698–2891>)	120	Ashraf et al. (2013)
		BOD_5 (128–142>)	120	
		Cl (123.8–127.7>)	5	
		Ni (0.44–0.65>)	0.07	
		As (0.06–0.07>)	0.01	
		Pb (0.04–0.08>)	0.01	
Mexicali (Mexico)	Ground-water ($mg\text{-}L^{-1}$)	COD (23.5–188>) BOD_5 (4.3–6.5)	120* 20*	Reyes-López et al. (2008)
		Na (600>)	200	
		SO_4^- (1000>)	300	

Note: Soil-contamination limits (Efrymson and Will, 1997), drinking-water limits (World Health Organization, 2004).

*Water release after waste-water treatment.

discussions were made on the norms, laws, and additionally, to the principles as prescribed by the government for appropriate disposals of hospital-wastes (Dwivedi et al., 2009). In order to obtain a satisfactory running of recycling-process in developing-countries, the environmental-education should be blend with economic-instruments like government-subsidies (Diaz and Otama, 2013). Pani et al. (2016) have made a study on three classes of wastes that are hazardous in nature like " MSWs, Hospital-Wastes, and EW." Different techniques for management of wastes were discussed that included landfills, biological-reprocessing, pyrolysis, etc. It was concluded of creating making coaching-programs' awareness in association with wastes and its management. Garnaik et al. (2016) have applied "PROMETHEE-II, TOPSIS, and VIKOR analysis" for creating the choices of the simplest suppliers supported with different criteria for "Lenitive-Pharmaceuticals." Satapathy et al. (2015) have revealed that the economy of the society are often enhanced by waste management. By proper utilization of the paper, gallons of water also as plenty of trees are often saved by people and if managed appropriately then a substantial decrease in greenhouse gases are often created leading to lesser pollution. Rao et al. (2003) have suggested of putting in bio-medical treatment facilities or providing treatment of wastes at sources for every hospitals generating bio-medical wastes. Peoples from every level of society usually visit the institutions like hospitals irrespective of ages, races, sex, colors, religions, or social-stations. Marram (2011) has revealed that a serious economic and environmental issue is represented by the right-management of solid wastes throughout the international-borders, and an increased cost for correct waste disposals has been a competitive-opportunity for recycling. The Karst systems are very complex and because of hydrological and geographical factors, they are available amongst the foremost delicate and imperil environments (Brinkmann and Parise, 2012). Peoples should make a robust-pledge to find out consonant-measure with nature. Many problems faced by human-beings with the environment are often easily resolved by changing in human-systems instead of environment-alterations. An experiment was conducted by Ganiron Jr. (2013), which focused on the effects recycled-bottles utilization as concrete-materials for mass-housing projects in USA. It was aimed at determining the properties like 'compressive-strength and modulus of elasticity by using recycled-bottles instead of concretes. Kumar et al. (2013) have reported of medical-aid to be significant for all citizen, and the generated wastes are a true life challenging-problem. The survey was carried out in hospitals in Mysore

(India), where a greater volume of wastes are generated. It was concluded that disposal-techniques are not perfect as the advanced and newer technology are remaining un-adopted. They recorded the typical generation of a variety of infectious waste-items per hospital unit area. Moreover, the segregation and transportation of recyclable-materials would cause volume reduction of wastes for final-disposals and healthy-environment. These measures are not of so much cost and are also very effective in reducing wastes. In their study, they discussed about various treatment-process for biomedical-wastes like autoclave-treatment, hydroclave-treatment, and microwave-treatment. Eventually, it was further concluded that biomedical wastes management cannot be successfully implemented unless the people are self-motivated, devoted, and co-operative. The principles for correct-management and proper-handling of biomedical-waste was summarized by Babu et al. (2009). Discussions were carried out on various countries taking biomedical-wastes as a significant issue for future. The topics like radioactive-materials, mercury-containing instruments and PVC-plastics were explained that are generated from hospitals. It was concluded that lack of awareness is the major cause for the problem for non-management of biomedical-wastes. Tandon et al. (2013) have made discussion about paper-industry controlling the economic-process. The environmental-load on manufacturing paper and additionally, the gathering-process of papers in India was also discussed. Nagavallemma et al. (2006) have made an investigation on vermicomposting. The nutrient-composition of vermicompost and garden compost were elaborated as manganese, iron, phosphorus, sodium, calcium, zinc, organic-carbon, etc. The development in yielding and growing of crops by the implementation of 'Vermicompost' was also discussed in their study. Sinha et al. (2010) have revealed of the entire elimination of the chemical fertilizers having harmful effect on soil, crops, and farmers, through earthworms and its' excreta. Khan and Ishaq (2011) have made a chemical-analysis for different composts like vermicomposts, pit-composts, and also, for garden-soils. The effect of those composts on crop-expansion called "*Pisum-sativum*" was found. It was concluded of richness of vermicomposts with nutrients like magnesium, calcium, chloride, potassium, nitrate, and sodium, which can improve the expansion of plants. They studied the expansion of plants in pots having vermicompost for one month period. Hai and Ali (2005) have studied the generation as well as characteristics of solid wastes in Dhaka (Bangladesh). They have analyzed the effects of the solid wastes on environment, and discussed about mechanism or practices for solid wastes management. According to their estimation, the wastes generation rate can cross 30 thousand tons per day in 2020, which is at present about 3500 tons per day. They concluded that with the assistance of composting, the speed of methane-gas generation will hamper to 50%, and also, in the near future, the transportation-value of wastes for dumping will increase owing to the unavailability of lands near the town. The Gazette of India (2010) has laid-down the responsibility of "municipal-authority, government, Union-Territory Administration, Central Pollution Control Board (CPCB), State Pollution Control Board" regarding wastes treatment and disposals. It was discussed regarding the compliance-criteria in view of collection, segregation, storage, transportation, processing, and disposals of solid wastes. It also laid down the specification for landfill-sites and the required facilities' availability at the location.

Whenever, a number of conflicting criteria exists, then the multi-criteria decision-making (MCDM) methods find their suitable applicability. For instance, Opricovnic and Tzeng (2004) have made a compromised solution by using MCDM methods like "Vlse Kriterijumska Optimizacija I Kompromisno Resenje (VIKOR)" and TOPSIS. Samantra et al. (2012) have applied the fuzzy-based VIKOR method to pick and evaluate an appropriate supplier for a supply chain management. With the use of PROMETHEE-II method (developed by Brans and Vincke, 1985), the choice of industrial-robot was made (Sen and Patel, 2015). By setting up a '3MW landfill gas-based power plant' on DBOOT basis for the conversion of wastes into energy at Gurai, Mumbai was proposed by Babu et al. (2009), with the suggestion about the barriers for waste management. Rawabdeh (2011) has presented a model by the utilization of "quality function deployment (QFD)" for the identification, prioritization, and determination of shop-floor wastes' sources in order to eliminate them. The barriers to reverse-logistics implementation in Polish enterprises was discussed in a study by Starostka-Patyk et al. (2014). Nikakhtar et al. (2015) have considered a construction process and explained about different kinds of wastes that can be reduced by adopting lean-construction principles with the use of computer simulation.

3.2 METHODOLOGY

To find the significant barriers in wastes management in India, an extensive literature review was done and suggestion were taken from experts of different areas such as "Municipal Corporation," "CPCB," and "Odisha State Pollution Control Board." Further, in order to rank the most important common-barriers, "VIKOR Analysis" was utilized (Figure 3.4).

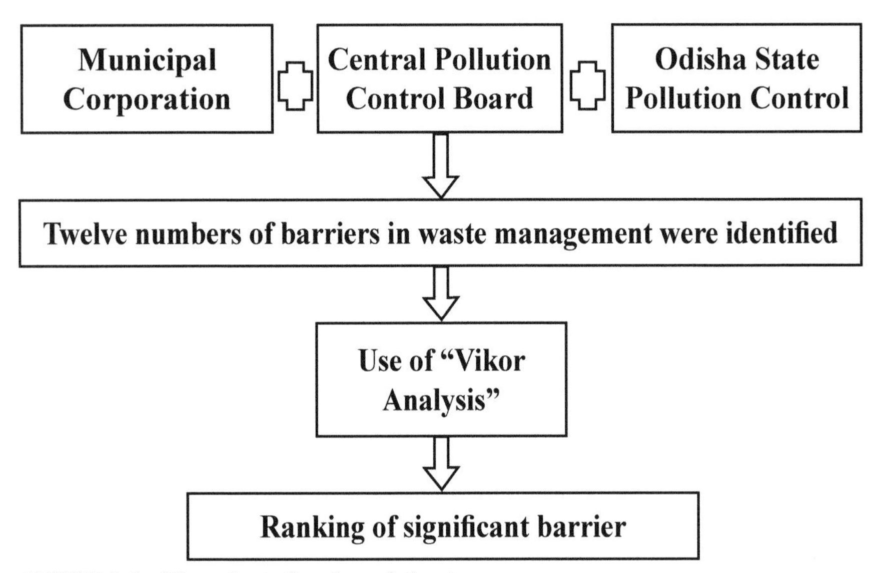

FIGURE 3.4 Flow-chart of work carried out.

3.3 RESULTS AND DISCUSSION

Based on the literature and discussion with experts,' the selected and suggested common-barriers for different types of wastes like hospital-wastes, MSWs and PW were:

- Budgets allocation by municipalities for solid wastes management;
- Bulk-expenditure on collection and transportation;
- Inadequate budgets on processing/treatment;
- Tremendous-processing and disposals of MSWs;
- Inadequate disposal-system;
- Inadequate financial and institutional capability;
- Not having viable business-module in the sector;
- Dependency of municipalities for budget on 'State and Central Government';
- Identifying suitable-sites;

- Compliance for waste-payment;
- Training and awareness programs;
- Requirement of appropriate technology.

Further, the "VIKOR Analysis" was used in order to rank and prioritize these barriers for wastes management. It may be noted that the "VIKOR Analysis" method is a "multi-criteria decision-making (MCDM)" or "multi-criteria decision analysis (MCDA)" method originally developed by "Serafim Opricovic" for solving decision-problems with conflicting-criteria, assuming that compromising is acceptable for conflict-resolution. The barriers are calculated as per the established-criteria and the decision-maker gets an inference which is closest to the ideal. This method ranks barriers and determines the compromise-solution which is the closest to the ideal. The following steps were followed in "VIKOR Analysis":

- **Step I:** The positive ideal-solutions (f_j^*) and negative ideal solutions (f_j^-) were obtained as follows:

$$f_j^* = \left\{ \max_{i=1,\ldots,m} f_{ij} \right.$$

$$f_j^- = \left\{ \min_{i=1,\ldots,m} f_{ij} \right.$$

where; j = 1, n.

- **Step II:** Values of "S_i and R_I" were calculated by using the following formulae:

$$S_i = \sum_{j=1}^{n} w_j (f_j^* - f_{ij}) / (f_j^* - f_j^-)$$

$$R_i = \max_{j=1,\ldots,n} [w_j (f_j^* - f_{ij}) / (f_j^* - f_j^-)]$$

- **Step III:** The values of "Q_i" were evaluated by the following equation:

$$Q_i = \{v (S_i - S^*)/(S^- - S^*)\} + \{(1-v)(R_i - R^*)/(R^- - R^*)\}$$

where; $S^* = \min_{i=1,\ldots,m} S_i$, $S^- = \max_{i=1,\ldots,n} S_i$.

$R^* = \min_{i=1,\ldots,m} R_i$, $R^- = \max_{i=1,\ldots,n} R_i$.

where; v is the weight taken to be 0.5.

- **Step IV:** The barriers were ranked by arranging "S, R, and Q" in ascending-order.

- **Step V:** If the following two statements were satisfied simultaneously, then the scheme with minimum value of Q in ranking was considered the optimal compromise-solution. Such as:
  - **Statement 1:** The barrier "$Q(A^{(1)})$" has an acceptable-advantage.

    $$Q(A^{(2)}) - Q(A^{(1)}) \geq 1/(m^{-1})$$

    where; "$A^{(2)}$" is the barrier with the second-position in the ranking-list and 'm' is the number of barriers.
  - **Statement 2:** The barrier "$Q(A^{(1)})$" is stable within the decision-making process. In other words, it is also best ranked in "S_i and R_I."

    If Statement 1 is not satisfied, that means "$Q(A^{(m)}) - Q(A^{(1)}) < 1/(m^{-1})$," then barriers "$A^{(1)}$, $A^{(2)}$, ..., $A^{(m)}$" all are the compromise-solution, there is no comparative-advantage of "$A^{(1)}$" from others. But, for the case of maximum-values, the corresponding barriers is the compromise-solution. If Statement 2 is not satisfied, the stability in decision-making is deficient while "$A^{(1)}$" has a comparative-advantage. Therefore, "$A^{(1)}$ and $A^{(2)}$" will have the same compromise-solution.
- **Step VI:** The barrier which is the most important to be looked upon was selected by choosing "$Q(A^{(m)})$" as a best compromise-solution with the minimum-value of "Q_i."

Tables 3.7–3.9 represented the numerical data for barriers selection, values of "S, R, and Q" for all barriers, and the barriers-ranking by "S, R, and Q" in ascending-order, respectively.

From "VIKOR analysis," the most important barriers found were as follows: "Training and awareness programs," "Bulk-expenditure on collection and transportation" and "Dependency of municipalities for budget on State and Central Government." These barriers not only will help to solve issues related to wastes management, but also will be helpful in framing-policies for proper management of wastes.

3.4 CONCLUSION

NGO's and diverse associations ought to create awareness among the individuals about waste disposals and its harmful effects on the

TABLE 3.7 Numerical Data for Barriers Selection

Barriers	Symbols-Assigned	Paper-Wastes	Municipal Solid Wastes	Hospital-Wastes
Budgets allocation by municipalities for solid wastes management	A	3.5	3.67	4
Bulk-expenditure on collection and transportation	B	3.17	2.67	3.33
Inadequate budgets on processing/treatment	C	3.17	2.83	3.83
Tremendous-processing and disposals of municipal solid wastes	D	3.67	3	3
Inadequate disposal-system	E	4.33	3	2.83
Inadequate financial and institutional capability	F	3.33	3.17	3.67
Not having viable business-module in the sector	G	3.67	3.67	3
Dependency of municipalities for budget on 'State and Central Government'	H	3.67	3.5	4
Identifying suitable-sites	I	3.17	3.5	3.16
Compliance for waste-payment	J	3.67	3.5	2.67
Training and awareness programs	K	4.67	3.5	4.33
Requirement of appropriate technology	L	4	3.33	3.33

TABLE 3.8 The Values of "S, R, and Q" for All Barriers

	A	B	C	D	E	F	G	H	I	J	K	L
S	2.74	9.58	7.62	8	7.31	6.15	4.87	3.25	6.11	6.46	0.79	5.11
R	1.95	4.67	3.92	3.20	3.61	2.33	3.20	1.67	2.81	4	0.79	2.41
Q	0.26	1	0.79	0.72	0.73	0.50	0.54	0.25	0.56	0.73	0	0.45

TABLE 3.9 The Barriers-Ranking by "S, R, and Q" in Ascending-Order

S-Ranking	**K**	**A**	**H**	**G**	**L**	**I**	**F**	**J**	**E**	**C**	**D**	**B**
S	0.79	2.74	3.25	4.87	5.11	6.11	6.15	6.46	7.31	7.62	8	9.58
R-Ranking	**K**	**H**	**A**	**F**	**L**	**I**	**G**	**D**	**E**	**C**	**J**	**B**
R	0.79	1.66	1.95	2.33	2.41	2.81	3.20	3.20	3.61	3.92	4	4.67
Q-Ranking	**K**	**H**	**A**	**L**	**F**	**G**	**I**	**D**	**E**	**J**	**C**	**B**
Q	0	0.25	0.26	0.45	0.50	0.54	0.56	0.72	0.73	0.73	0.79	1
Overall-Ranking	1	3	4	5	5	6	7	8	9	9	10	2

environment as well as on their well-being. By utilizing the multi-criteria decision-making methods, the most important barriers that need appropriate attention were obtained as "Training and awareness programs," followed by "Budgets allocation by municipalities for solid wastes management" and "Dependency of municipalities for budget on State and Central Government," and so on. The consequence of this examination will provide an answer to the administration for improving in the territories, wherein they need proper wastes disposals and management. This may lead to resolve the problems associated with the waste management and will help the peoples to get awareness of the harmful-effects of wastes on their health and the surroundings. Awareness and training programs will make a positive impact on the individuals, and they would likewise begin lessening the creation of wastes. This research gives the knowledge to outline new policies against waste management, but there may exist cases of fluctuating number of barriers by a greater number of experts' analysis information.

KEYWORDS

- **anaerobic-digestion**
- **bio-drying**
- **civil solid wastes**
- **clinic or bio-clinical wastes**
- **e-wastes**
- **paper-wastes**
- **waste management**

REFERENCES

Aderoju, O. M., Dias, G. A., & Gonçalves, A. J., (2018). A GIS-based analysis for sanitary landfill sites in Abuja, Nigeria. *Environ. Dev. Sustain.*, 1–24.

Ali, M., Wang, W., Chaudhry, N., & Geng, Y., (2017). Hospital waste management in developing countries: A mini-review. *Waste Manag. Res., 35,* 581–592.

Al-Khatib, I. A., Abu, H. A., Sharkas, O. A., & Sato, C., (2015). Public concerns about and perceptions of solid waste dump sites and selection of sanitary landfill sites in the West Bank, Palestinian territory. *Environ. Monit. Assess, 187,* 186.

Ashraf, M. A., Yusoff, I., Yusof, M., & Alias, Y., (2013). Study of contaminant transport at an open-tipping waste disposal site. *Environ. Sci. Pollut. Res., 20,* 4689–4710.

Babu, B. R., Parande, A. K., Rajalakshmi, R., Suriyakala, P., & Volga, M., (2009). Management of biomedical waste in India and other countries: A review. *J. Int. Environmental Application and Science, 4*(1), 65–78.

Babu, S. S., Chakrabarti, T., Bhattacharyya, J. K., Biswas, A. K., & Kumar, S., (2010). Studies on environmental quality in and around municipal solid waste dumpsite. *Resour. Conserv. Recycle., 55,* 129–134.

Bernardes, A. M., Espinosa, D. C. R., & Tenório, J. A. S., (2004). Recycling of batteries: A review of current processes and technologies. *J. Power Sources, 130,* 291–298.

Betts, K., (2008). Producing usable materials from e-waste, *Environ. Sci. Technol., 42,* 6782–6783.

Bing, X., Bloemhof, J. M., Ramos, T. R. P., Barbosa-Povoa, A. P., Wong, C. Y., & Van, D. V. J. G. A. J., (2016). Research challenges in municipal solid waste logistics management. *Waste Manag., 48,* 584–592.

Brans, J. P., & Vincke, P. H., (1985). A preference ranking organization method: The PROMETHEE method for multiple criteria decision-making. *Management Science, 31*(6), 647–656.

Brinkmann, R., & Parise, M., (2012). Karst environments: Problems, management, human impacts, and sustainability: An introduction to the special issue. *Journal of Cave and Karst Studies.* doi: 10.4311/2011JCKS0253.

Brown, D., & McGranahan, G., (2016). The urban informal economy, local inclusion and achieving a global green transformation. *Habitat Int., 53,* 97–105.

Chiemchaisri, C., Juanga, J. P., & Visvanathan, C., (2007). Municipal solid waste management in Thailand and disposal emission inventory. *Environ. Monit. Assess., 135,* 13–20.

Cobbing, M., (2008). *Toxic Tech: Not in Our Backyard.* Uncovering the hidden flows of e-waste. Report from Greenpeace International. Amsterdam.

Díaz, R. O., Hernández, M. M., Echeverría, C. F., & Arado, L. J. O., (2012). Assessment of metal pollution in soils from a former Havana (Cuba) solid waste open dump. *Bull. Environ. Contam. Toxicol., 88,* 182–186.

Diaz, R., & Otoma, S., (2013). Constrained recycling: A framework to reduce landfilling in developing countries. *Waste Management and Research, 31*(1), 23–29.

Dos, M. L. S., Tokai, A., & Hanashima, A., (2015). Analyzing the structure of barriers to municipal solid waste management policy planning in Maputo city, Mozambique. *Environ. Dev., 16,* 76–89.

Dwivedi, A. K., Pandey, S., & Shashi, (2009). Hospital waste: At a glance. In: Trivedi, P. C., (ed.), *Microbes Applications and Effect* (pp. 114–119). Aavishkar Publishers and Distributors, Jaipur, India. (ISBN: 978-81-7910271-8).

Efrymson, M. E., & Will, G. W. S., (1997). Toxicological benchmarks for contaminants of potential concern for effects on soil and litter invertebrates and heterotrophic process. *Ecotoxicol. Environ. Saf.*

Ezeah, C., Fazakerley, J. A., & Roberts, C. L., (2013). Emerging trends in informal sector recycling in developing and transition countries. *Waste Manag., 33,* 2509–2519.

Ferronato, N., Rada, E. C., Gorritty, P. M. A., Cioca, L. I., Ragazzi, M., & Torretta, V., (2019). Introduction of the circular economy within developing regions: A comparative analysis of advantages and opportunities for waste valorization. *J. Environ. Manag., 230,* 366–378.

Ferronato, N., Torretta, V., Ragazzi, M., & Rada, E. C., (2017). Waste mismanagement in developing countries: A case study of environmental contamination. *UPB Sci. Bull. Ser. D Mech. Eng., 79,* 185–196.

Gaidajis, G., Angelakoglou, K., & Aktsoglou, D., (2010). E-waste: Environmental problems and current management. *Journal of Engineering Science and Technology Review, 3*(1), 193–199.

Ganiron, Jr. T. U., (2013). Use of recycled glass bottles as fine aggregates in concrete mixture. *International Journal of Advanced Science and Technology, 61,* 17–28.

Garnaik, A., Kumar, S., Ray, S. K., Majumder, A., & Pani, A., (2016). Use of multi-criteria decision-making tools in supplier selection for lenitive pharmaceuticals: A case study. *International Journal of Research and Scientific Innovation, 3*(II), 5–9.

Ghisolfi, V., Chaves, G. D. L. D., Siman, R. R., & Xavier, L. H., (2017). System dynamics applied to closed-loop supply chains of desktops and laptops in Brazil: A perspective for social inclusion of waste pickers. *Waste Manag, 60,* 14–31.

Grossman, E., (2006). *High Tech Trash: Digital Devices, Hidden Toxics, and Human Health.* Washington: Island Press/Shearwater Books.

Gupta, N., Yadav, K. K., & Kumar, V., (2015). A review on current status of municipal solid waste management in India. *J. Environ. Sci. (China), 37,* 206–217.
Gutberlet, J., & Baeder, A. M., (2008). Informal recycling and occupational health in Santo André, Brazil. *Int. J. Environ. Health Res., 18,* 1–15.
Gutberlet, J., Baeder, A. M., Pontuschka, N. N., Felipone, S. M. N., & Dos, S. T. L. F., (2013). Participatory research revealing the work and occupational health hazards of cooperative recyclers in Brazil. *Int. J. Environ. Res. Public Health, 10,* 4607–4627.
Hai, F. I., & Ali, M. A., (2005). A Study on solid waste management system of Dhaka city corporation: Effect of composting and landfill location. *UAP Journal of Civil and Environmental Engineering, 1*(1).
Hettiarachchi, H., Meegoda, J. N., & Ryu, S., (2018). Organic waste buyback as a viable method to enhance sustainable municipal solid waste management in developing countries. *Int. J. Environ. Res. Public Health, 15,* 2483.
Ihedioha, J. N., Ukoha, P. O., & Ekere, N. R., (2017). Ecological and human health risk assessment of heavy metal contamination in soil of a municipal solid waste dump in Uyo, Nigeria. *Environ. Geochem. Health, 39,* 497–515.
Imam, A., Mohammed, B., Wilson, D. C., & Cheeseman, C. R., (2008). Solid waste management in Abuja, Nigeria. *Waste Manag., 28,* 468–472.
Inglezakis, V. J., & Moustakas, K., (2015). Household hazardous waste management: A review. *J. Environ. Manag., 150,* 310–321.
Kanmani, S., & Gandhimathi, R., (2012). Assessment of heavy metal contamination in soil due to leachate migration from an open dumping site. *Appl. Water Sci., 3,* 193–205.
Karak, T., Bhattacharyya, P., Das, T., Paul, R. K., & Bezbaruah, R., (2013). Non-segregated municipal solid waste in an open dumping ground: A potential contaminant in relation to environmental health. *Int. J. Environ. Sci. Technol., 10,* 503–518.
Khan, A., & Ishaq, F., (2011). Chemical nutrient analysis of different composts (vermicompost and pit compost) and their effect on the growth of a vegetative crop *Pisum sativum. Asian J. Plant Sci. Res., 1*(1), 116–130.
Kumar, V., Yadav, K., & Rajamani, V., (2013). Selection of Suitable site for solid waste management in part of Lucknow City, Uttar Pradesh using remote sensing, GIS and A.H.P method. *International Journal of Engineering Research and Technology, 2*(9).
Li, J. H., Gao, S., Duan, H. B., & Liu, L. L., (2009). Recovery of valuable materials from waste liquid crystal display panel. *Waste Manag., 29,* 2033–2039.
Lim, S. L., Lee, L. H., & Wu, T. Y., (2016). Sustainability of using composting and vermicomposting technologies for organic solid waste biotransformation: recent overview, greenhouse gases emissions and economic analysis. *J. Clean. Prod., 111,* 262–278.
Linzner, R., & Salhofer, S., (2014). Municipal solid waste recycling and the significance of informal sector in urban China. *Waste Manag. Res., 32,* 896–907.
Lohri, C. R., Rajabu, H. M., Sweeney, D. J., & Zurbrügg, C., (2016). Char fuel production in developing countries: A review of urban biowaste carbonization. *Renew. Sustain. Energy Rev., 59,* 1514–1530.
Magram, S. F., (2011). *Indian Journal of Science and Technology, 4*(6).
Manaf, L. A., Samah, M. A. A., & Zukki, N. I. M., (2009). Municipal solid waste management in Malaysia: Practices and challenges. *Waste Manag., 29,* 2902–2906.
Matter, A., Ahsan, M., Marbach, M., & Zurbrügg, C., (2015). Impacts of policy and market incentives for solid waste recycling in Dhaka, Bangladesh. *Waste Manag., 39,* 321–328.

Modak, P., Wilson, D. C., & Velis, C., (2015). Waste management: Global status. In: *Global Waste Management Outlook* (pp. 51–79). UNEP: Athens, Greece. ISBN: 9789280734799.

Nagavallemma, K. P., Wani, S. P., Stephane, L., Padmaja, V. V., Vineela, C., Babu, R. M., & Sahrawat, K. L., (2006). Vermicomposting: Recycling wastes into valuable organic fertilizer. *ICRISAT, 2*(1).

Nikakhtar, A., Hosseini, A. A., Wong, K. Y., & Zavichi, A., (2015). Application of lean construction principles to reduce construction process waste using computer simulation: A case study. *International Journal of Services and Operations Management, 20*(4), 461–480.

Ongondo, F. O., Williams, I. D., & Cherrett, T. J., (2011). How are WEEE doing? A global review of the management of electrical and electronic wastes. *Waste Manag., 31,* 714–730.

Opricovic, S., & Tzeng, G. H., (2004). Compromise solution by MCDM methods: A comparative analysis of VIKOR and TOPSIS. *European Journal of Operational Research, 156*(2), 445–455.

Ouda, O. K. M., Raza, S. A., Nizami, A. S., Rehan, M., Al-Waked, R., & Korres, N. E., (2016). Waste to energy potential: A case study of Saudi Arabia. *Renew. Sustain. Energy Rev., 61,* 328–340.

Pani, A., Garnaik, A., & Swarn, A., (2016). Governance over wastes all-round the universe. *Proceedings of National Conference on Recent Trends in Environment, Science and Technology*, 131–135.

Prechthai, T., Parkpian, P., & Visvanathan, C., (2008). Assessment of heavy metal contamination and its mobilization from municipal solid waste open dumping site. *J. Hazard. Mater., 156,* 86–94.

Rao, S. K. M., Ranyal, R. K., Bhatia, S. S., & Sharma, V. R., (2003). Biomedical waste management: An infrastructural survey of hospitals. *MJAFI, 60*(4).

Rawabdeh, I., (2011). Waste elimination using quality function deployment. *International Journal of Services and Operations Management, 10*(2), 216–238.

Reyes-López, J. A., Ramírez-Hernández, J., Lázaro-Mancilla, O., Carreón-Diazconti, C., & Garrido, M. M. L., (2008). Assessment of groundwater contamination by landfill leachate: A case in México. *Waste Manag., 28,* S33–S39.

Robinson, B., (2009). E-waste: An assessment of global production and environmental impacts. *Science of the Total Environment, 408,* 183–191.

Sadef, Y., Nizami, A. S., Batool, S. A., Chaudary, M. N., Ouda, O. K. M., Asam, Z. Z., Habib, K., et al., (2016). Waste-to-energy and recycling value for developing integrated solid waste management plan in Lahore. *Energy Sources Part B Econ. Plan Policy, 11,* 569–579.

Samantra, C., Datta, S., & Mahapatra, S. S., (2012). Application of fuzzy based VIKOR approach for multi-attribute group decision making (MAGDM): A case study in supplier selection. *Decision Making in Manufacturing and Services, 6*(1), 25–39.

Sanneh, E. S., Hu, A. H., Chang, Y. M., & Sanyang, E., (2011). Introduction of a recycling system for sustainable municipal solid waste management: A case study on the greater Banjul area of The Gambia. *Environ. Dev. Sustain., 13,* 1065–1080.

Sasaki, S., & Araki, T., (2014). Estimating the possible range of recycling rates achieved by dump waste pickers: The case of Bantar Gebang in Indonesia. *Waste Manag. Res., 32,* 474–481.

Satapathy, S., Kumar, S., & Garnaik, A., (2016). Eradicating the barriers: Betterment of waste management. *International Journal of Research and Scientific Innovation, 3*(I), 41–44.

Sawadogo, M., Tchini, T. S., Sidibé, S., Kpai, N., & Tankoano, I., (2018). Cleaner production in Burkina Faso: Case study of fuel briquettes made from cashew industry waste. *J. Clean. Prod., 195,* 1047–1056.

Sen, D., & Patel, S. K., (2015). Multi-criteria decision making towards selection of industrial robot. *Benchmarking: An International Journal, 22*(3), 465–487.

Seng, B., Fujiwara, T., & Spoann, V., (2018). Households' knowledge, attitudes, and practices toward solid waste management in suburbs of Phnom Penh, Cambodia. *Waste Manag. Res., 36,* 993–1000.

Singh, M., Thind, P. S., & John, S., (2018). Health risk assessment of the workers exposed to the heavy metals in e-waste recycling sites of Chandigarh and Ludhiana, Punjab, India. *Chemosphere, 203,* 426–433.

Sinha, R. K., Agarwal, S., Chauhan, K., & Valani, D., (2010). The wonders of earthworms and its vermicompost in farm production: Charles Darwin'sfriends of farmers,' with potential to replace destructive chemical fertilizers from agriculture. *Agricultural Sciences, 1*(2), 76–94. doi: 10.4236/as.2010.12011.

Starostka-Patyk, M., Zawada, M., Pabian, A., & Szajt, M., (2014). *Reverse Logistics Barriers in Polish Enterprises, 19*(2).

Sustainable Development Goals, (2016). *Fund Goal 12: Responsible Consumption and Production.* United Nations Dev. Program, SDG Fund: New York, NY, USA.

Tandon, R., Negi, S. D., & Mathur, R. M., (2013). *Waste Paper Collection Mechanism in India-Current Status and Future Requirement* (Vol. 25, No. 3). IPPTA.

The World Bank, (2012). *What a Waste: A Global Review of Solid Waste Management.* The World Bank: Washington, DC, USA.

Thi, N. B. D., Kumar, G., & Lin, C. Y., (2015). An overview of food waste management in developing countries: Current status and future perspective. *J. Environ. Manag., 157,* 220–229.

Tian, H., Gao, J., Hao, J., Lu, L., Zhu, C., & Qiu, P., (2013). Atmospheric pollution problems and control proposals associated with solid waste management in China: A review. *J. Hazard. Mater., 252,* 152–154.

Torretta, V., Ferronato, N., Katsoyiannis, I. A., Tolkou, A. K., & Airoldi, M., (2017). Novel and conventional technologies for landfill leachates treatment: A review. *Sustaining., 9,* 9.

UN, (2018). *The Sustainable Development Goals Report.* UN: New York, NY, USA.

UNEP (2013). *Guidelines for National Waste Management Strategies: Moving from Challenges to Opportunities.* Retrieved from: https://cwm.unitar.org/national-profiles/publications/cw/wm/UNEP_UNITAR_NWMS_English.pdf (accessed on 18 February 2021).

UNEP, (2011). *The Green Economy Report: A Preview.* Retrieved from: https://all62.jp/ecoacademy/images/15/green_economy_report.pdf (accessed on 18 February 2021).

UNEP-Global, (2009). *Marine Litter: A Global Challenge.* UNEP: Athens, Greece. ISBN: 9789280730296.

Vaccari, M., Vinti, G., & Tudor, T., (2018). An analysis of the risk posed by leachate from dumpsites in developing countries. *Environments, 5,* 99.

Vitorino, D. S. M. A., Montenegro, G. S., Faceli, K., & Casadei, V., (2017). Technologies and decision support systems to aid solid-waste management: A systematic review. *Waste Manag., 59,* 567–584.

Vongdala, N., Tran, H. D., Xuan, T. D., Teschke, R., & Khanh, T. D., (2019). Heavy metal accumulation in water, soil, and plants of municipal solid waste landfill in Vientiane, Laos. *Int. J. Environ. Res. Public Health, 16,* 22.

Wiedinmyer, C., Yokelson, R. J., & Gullett, B. K., (2014). Global emissions of trace gases, particulate matter, and hazardous air pollutants from open burning of domestic waste. *Environ. Sci. Technol., 48*, 9523–9530.

Wilson, D. C., & Velis, C. A., (2015). Waste management-still a global challenge in the 21st century: An evidence-based call for action. *Waste Manag. Res.,* 1049–1051.

World Bank, (2012). *What a Waste: A Global Review of Solid Waste Management*. Urban Development Series Knowledge Papers. Retrieved from: http://web.world-bank.org (accessed on 18 February 2021).

World Health Organization, (2004). *Guidelines for Drinking-Water Quality* (3rd edn.). WHO Libr. Cat. Data: Geneva, Switzerland.

Yukalang, N., Clarke, B., & Ross, K., (2017). Barriers to effective municipal solid waste management in a rapidly urbanizing area in Thailand. *Int. J. Environ. Res. Public Health, 14,* 1013.

CHAPTER 4

A FUZZY-COPRAS APPROACH FOR EFFECTIVE AGRICULTURAL WASTE MANAGEMENT

ABSTRACT

Agricultural improvement is normally accompanied by wastes from the unreasonable use of farming techniques intensively and the ill-treatment of synthetic substances in cultivations influencing the rural as well as the global environment. The wastes generated depend primarily on the system and type of farming activities. Usually, the agricultural wastes produced from cultivation are currently used for several applications through appropriate waste management. In this study, the farm waste management system in the Indian context was discussed, and a fuzzy-COPRAS approach for effective waste management was used.

4.1 INTRODUCTION

Agricultural wastes mostly include the residues from the growth and processing of raw agricultural products like vegetables, fruits, crops, poultry, meat, and dairy products. The composition of these wastes depends on the system as well as the type of agricultural activities that can be in the form of solids, liquids, or slurries. Although there is a rare estimation for the agricultural wastes, but these are normally considered to be contributing a significant proportion of the total waste matters in the world. Agricultural waste management is becoming a complex-domain with the interaction of several dimensions. Thus, the decision-makers are imposed with continuous challenges for the analysis and control of agricultural wastes. In this context, the multicriteria decision-making (MCDM) models have become significant and helpful supporting tools for waste management as these are able to handle a variety of problems involving several dimensions as well as conflicting criteria. Moreover, because of the availability of several multicriteria decision-making approaches, it becomes a hard task in the selection of the multicriteria decision-making method. Thus, in order to support the decision-makers and researchers, the objective of this study was to propose an effective waste management in India.

Usually after using the pesticides by farmers, most of the packages and bottles carrying these pesticides are thrown into ponds or fields. Most of these wastes have the potential of causing un-predictable environmental consequences, for example, poisoning of foods, un-safe food hygiene and contaminated farm-land owing to their potential lasting as well as toxic chemicals. Figure 4.1 illustrates the agricultural vegetable wastes that were disposed off without any precautionary actions. The Plant Protection Department (PPD) has made an estimation that about 1.8% of the chemicals remains in their packages (Dien and Vong, 2006). Serious damages on the environment and population are caused by these materials, especially when in spite of eliminating these are left on water supplies, sewage systems, collecting, and storage areas of garbage. Therefore, transportation and elimination of solid wastes is a must after collecting them in the proper technical and medical conditions (Muşdal, 2007). Tekel (2007) have classified the solid waste management system as integrated in addition to sustainable solid waste management systems. Ohman et al. (2007) have used the analytic hierarchical process (AHP) method in order to evaluate the alternatives for solid waste disposals. Khan and Faisal (2008) have proposed a methodology to prioritize the urban solid-waste disposal methods at local municipality to select the appropriate one by decision-makers, through the use of an ANP process with five main criteria in addition to thirteen sub-criteria for evaluating three alternatives. The expanding agricultural production has resulted in increased amounts of livestock wastes, agricultural crop-residues in addition to agro-industrial by-products. Thus, if developing countries continue to intensify farming systems, then there is likely to be a significant increase in agricultural wastes worldwide. About 998 million tons of agricultural wastes has been estimated to be produced yearly (Agamuthu, 2009). In our day-to-day activities we produce the solid wastes comprising of all materials such as bottles, package, leftovers, equipment, newspapers, devices, dyes, and batteries, etc. (Ekmekçioğlu et al., 2010). Batarseh et al. (2010) have made an investigation of sustainable urban solid wastes disposal by focusing on bio-reactor storages. The disordered storage and disposal of solid wastes lead to the occurrence of several problems such as quick spreading of infectious diseases, soil contaminations, surface water pollution, groundwater contaminations, irritating odors emission to the surroundings, pest, and insect problems, landslides, explosions, air pollution, etc. (Büyükbektaş and Varınca, 2010). Generally, higher productivity can be achieved through the use of inorganic and inexpensive fertilizers. However, more fertilizers are applied by many farmers than the amount needed to crops (Hai and Tuyet, 2010). The percentage of NH_3, CH_4, and H_2S varies in conjunction with the phases of the digestion progression and also depends on food components, organic materials, micro-organisms, and animals' health status. The greenhouse gases can be generated by these un-treated and non-reusable waste sources having negative effects on the soil fertility in addition to resulting in water pollutions. In livestock wastes, the water volume ranges between 75 to 95% of total volume with organic matters, in-organic matters, and several species of micro-organisms as well as parasite eggs as residues (Hai and Tuyet, 2010). Tezçakar and Can (2011) have explained the thermal disposal technology in order to create energy from wastes. Pires et al. (2011) have assessed the sustainable growth of solid waste-management system in Setúbal Peninsula regions (Portugal), by using 4 main criteria, 14 sub-criteria and 5 alternatives by the use of analytic network process (ANP) based on TOPSIS method. An integrated management program with due emphasis on economic, political, environmental, technical, and social-cultural components leads to a sustainable solid waste management system (Zurbrügg et al., 2012). Eskandari et al. (2012) have made an attempt to select the best place for solid waste-disposal in context with the environmental, economic as well as social-cultural visions for Marvdasht region (Iran), by using an integrated multicriteria decision-making approach. Similarly, Hanan et al. (2012) have also used multicriteria decision analysis to evaluate the waste paper management in Isle (Wight-island in UK), by utilizing seven recycling, recovery as well as disposal alternatives in terms of social and environmental criteria. Karmperis et al. (2012) have used a risk-based multicriteria evaluation approach for project alternatives in waste management.

Nowadays, the rapid increase in population as well as the changing life standards makes it difficult to manage and control the waste compositions,

FIGURE 4.1 Disposed agricultural vegetables wastes.

volumes, and proper disposals. As the by-products of agricultural-activities are not the primary products, so these are usually referred to as "agricultural-wastes" that mainly take the form of crop-residues (residual-stalks, straws, roots, leaves, shells-etcetera, husks) and animal-wastes or manures. These agricultural-wastes are commonly available, virtually-free, and renewable, thus making them as important resources (Sabiiti et al., 2005). There has been a greater requirement of awareness among the farmers and public for the benefits of organic-wastes' utilization and their proper-management in agriculture in order to lessen preconceived-notions and fears of nuisance-problems, environmental-degradations as well as the decrement in land-values (Westerman and Bicudo, 2005). The agricultural wastes' impact on the environment not only depends on the generation amounts, but also on the used disposal-methods. The environment is polluted through some disposal-practices (Sabiiti et al., 2004; Tumuhairwe et al., 2009). For instance, agricultural wastes burning have been a usual practice in the under-developed countries, but act as a source of atmospheric-pollution. Ezcurra et al. (2001) have revealed that burning agricultural wastes release pollutants like carbon-monoxide, nitrous-oxide, nitrogen-dioxide, and particles (smoke-carbon), respectively. The formation of ozone and nitric-acid are accompanied by these pollutants (Hegg et al., 1987), which contribute to acid-depositions by posing risks to ecological as well as human health (Lacaux et al., 1992). Pollution causing odor was reported of contributing higher social-tensions among urban livestock-farmers in Kampala of Uganda (Katongole, 2009). The excessive animal-wastes application on land as fertilizers and soil-conditioners cause surface run-offs and leaching contaminating the surface or ground waters. As a result, nitrate-leaching has been regarded as a major 'nitrogen (N)' pollution-concern on livestock-farms (Mackie et al., 1998). Manure-decomposition can be a chief source of ammonia (NH_3), methane (CH_4) and nitrogen-oxides contributing in the accumulation of greenhouse gases. Ammonia-volatilization causes acid-deposition, which results in acid-precipitations (Lowe, 1995). Ozone-depletion is contributed due to emission of nitrous-oxides (N_2O) during the 'nitrification-denitrification cycle' (Schulte, 1997). As the agricultural wastes contain large amounts of organic-matters, and these can be used in enhancing the food-security by using as bio-fertilizers and soil-amendments, animal-feeds, and energy-production. In view of this, waste treatment technologies play a vital role in the improvement of soil fertility as well as the crop-productivity (Amoding, 2007; Hargreaves et al., 2008). In a study, it was revealed of importance of the use of organic-fertilizers in most parts of Africa, where lesser availabilities of nutrients have been a serious-constraint for food-production (Brouwer and Powell, 1998). Moreover, the process of composting also helps to reduce the volume of the wastes, thus it solves serious environmental issues with regard to the disposals of larger quantity of wastes, killing of present pathogens, decreasing in the weeds-germination in agricultural-fields, and in reducing odors (Jakobsen, 1995). The ruminants are found to be useful in converting crop-residues into foods that help to reduce the potential-pollutants. However, the rumens contain the microbial enzyme-cellulases, which are the only enzymes in digesting the most abundant plant-products, celluloses (CAST, 1975). The waste-disposal problems do not occur with ruminants, as nutrients in by-products are utilized (Oltjen and Beckett, 1996). Digestates are produced as the most important residues from anaerobic-digestions (ADs). These residues are extensively used in land-applications owing to their capabilities in improving the nutrients soil-retentions (Yang et al., 2015). Moreover, biomasses can be utilized in producing heat and electricity through different thermo-chemical technologies like gasification or combustions (Rincon et al., 2013). The stages in association with the crop-developments such as seeding, maintenance, harvesting, post-harvesting, industrial-transformations, etc., produce various quantities of residues that can be utilized as raw materials for the production of a broader range of products (Alzate et al., 2004; Lennartsson et al., 2014; Moncada et al., 2014; Pinzi and Dorado, 2012). Rice-husks are the non-edible agricultural-residues generated during the whole grain-dehusking phase, and it has been reported of 0.23 tons of rice-husks production from each ton of processed-rice (Kumar and Bandyopadhyay, 2006; Soltani et al., 2014). Based on the absence of enzymatic-capability for digestions, there are some restrictions in feeding-uses of these residues for ruminants in addition to other animals (Jeetah et al., 2015). Different proposals for valorization of rice-husks exist as concrete, ethanol, ceramics, adsorbents, agglomerates, and energy (Prada and Corte's, 2010). When applied in the average amount of 8 kg/m^2, the rice-husk residues can be used as fertilizers (Quiceno Villada and Mosquera Gutierrez, 2010). With larger structures having added dilute contents, the organisms that occur naturally convert manure organic-matters into carbon-dioxides and methane, and transform the organic-nutrients into plant-available mineral-forms (Burns et al., 1990). An effluent-irrigation requires a careful balancing of pond-prevention from the application-fields with the proper-timing as well as nutrient-application rates. Although, the reduction of odors is the consequence of anaerobic-treatment and also, the odors from lagoon-effluents are not so intense as that of slurry-manures, the emissions of odors and neighbor-complaints have been a problem for producers who use the lagoon-treatment systems (Lim et al., 2003; Mukhtar et al., 2010).

Day by day, the pollutions due to solid wastes and the number of potential risks associated with solid wastes are increasing; therefore, waste management has gained importance and becoming more complex. Accordingly, an integrated system consisting of all levels of waste management factors as well as relationships becomes necessary (Menikpura et al., 2013). It may be noted that since the late 1960s, decision support models have been developed in the field of waste management (Karmperis et al., 2013). The most widely used decision support frameworks include life cycle assessment (LCA), cost-benefit analysis, and multicriteria decision-making (MCDM) (Karmperis et al., 2013; Milutinović et al., 2014; Morrissey and Browne, 2004). By using LCA, an investigation was made on a less impactful municipal solid waste (MSW) management system. It was revealed that as municipalities can directly compare the actual environmental impacts of different planning options as well as technologies, hence LCA can be a useful tool for the planning of municipal waste management (Yay, 2015). The environmental aspects are mainly focused in LCA, and the cost-benefit analysis focuses on maximization of economic efficiency. Whereas, MCDM allows the consideration of economic, social, and environmental criteria as the three pillars of sustainability. MCDM methods have been considered as some of the most thorough as well as effective decision support frameworks for decision-making in solid wastes management (Soltani et al., 2015). Mardani et al. (2015) have revealed the energy, environment as well as sustainability as the first areas having the application of MCDM approaches and techniques. Obi et al. (2016) have discussed about the effects of the toxic agricultural wastes on the environment as well as the management of those toxic wastes. Kharat et al. (2016) applied the Fuzzy-AHP-TOPSIS based methodology to the MSW landfill site selection problem in Mumbai city (India), and suggested it as a powerful tool to guide the planners in MSW management. Mostly the problems associated with MSW that involve facility locations or management strategies have been addressed by different studies using multicriteria decision making in solid waste management (Coelho et al., 2017). The energy generation potential from landfills by means of thermal treatment or methane extraction is a major opportunity in India, but having a key barrier of qualified engineers' shortages as well as experienced environmental professionals for delivering improved systems of waste management (Kumar et al., 2017). Debnath et al. (2017) have used a hybrid MCDM approach by combining modified grey DEMATEL, multi-attributive border approximation-area comparison (MABAC), and sensitivity analysis, in order to evaluate as well as select the genetically modified (GM) agro-portfolio management, where the best portfolio was found to be the GM by-products. Through solid-state fermentation, the manufacturing of biofuels, vitamins, enzymes, antibiotics, animal feeds, antioxidants, and other chemicals are carried out by the agro-industrial wastes with the use of a variety of micro-organisms. Sadh et al. (2018) have reviewed and discussed about the solid-state fermentation along with their effects on the value-added products' formations. Si et al. (2018) have reviewed a total of 346 papers that were published between the years 2006 to 2016 about the methodologies as well as applications of the "decision making trial and evaluation laboratory (DEMATEL)" method, which is considered as an effective method for the cause-effect chain components identification of a complex system. The Indian government in addition to the governments of other developing countries can get advantages from the rising nexus thinking concept in order to manage the environmental resources, which promotes a higher level of involvement of stakeholders and higher-level integration going beyond the disciplinary boundaries, and providing a supporting platform to solve issues like burning of crop residues (Bhuvaneshwari et al., 2019). In order to minimize the nutrient gap, the recycling of surplus crops, horticultural residues, and animal excreta can be a viable option. However, there occurs a scarcity of cumulative estimated data on the availability of unutilized crops, horticultural residues, and animal excreta in India. Bhattacharjya et al. (2019) have made an attempt to give an estimation of recyclable bio-wastes in the states of central as well as western India useful for future policy-makers. Through many direct and indirect channels, the agricultural wastes have a toxicity potential to plants, animals, and human. Alkaradaghi et al. (2019) have proposed multicriteria decision-making (MCDM) methods for decision regarding landfill site in the north of Iraq, assuming two input groups of factors required for the optimal weight coefficients' values satisfaction. These groups of constants were natural and artificial factors, including 13 selected criterion. In addition to multicriteria decision-making (MCDM) methods, the weighted linear combination (WLC) method was used to derive criteria weights using a pair-wise comparison matrix.

4.2 METHODOLOGY

For the sustainability of agriculture with proper management of its wastes, the sustainability parameters such as economic, social as well as environmental impacts of the agricultural wastes that were considered as criterion by eight experts' opinions and literature, needed to be maintained to a minimum level. Moreover, as noted by USDA (2012), there exist six basic functions in the management of agricultural wastes (Figure 4.2) that includes production, collection, transport, storage, treatment, and utilization. Where, production refers to the amount as well as nature of agricultural wastes generated by agricultural sectors. If the quantity produced becomes sufficient, then these agricultural wastes need proper management so that these wastes turn into a resource concern. The complete production analysis includes the type or variety, consistency, amount, location, and waste production timings. Collection refers to the gathering of wastes from the origin point or deposition to a collecting point. Transfer refers to the movement of wastes all over the system. The temporary containment of the wastes is referred as storage. Treatment is designed for reducing the pollution potential or modifying the physical characteristics of the wastes in order to facilitate more effective and efficient handling. Whereas utilization refers to the reuse or recycle of wastes, as these agricultural wastes may be used as a source of energy, organic matters, or nutrients for plants.

Moreover, these functions in the management of agricultural wastes were considered as six alternatives, i.e., A_1, A_2, A_3, A_4, A_5, and A_6, as evaluated by the experts (Table 4.1). Further, as there may be subjective judgments' differences in the ways of experts' view in evaluating each criterion, hence in order to accomplish a rational and objective evaluation, the overall evaluation of the fuzzy judgment was employed for synthesizing the experts' opinions.

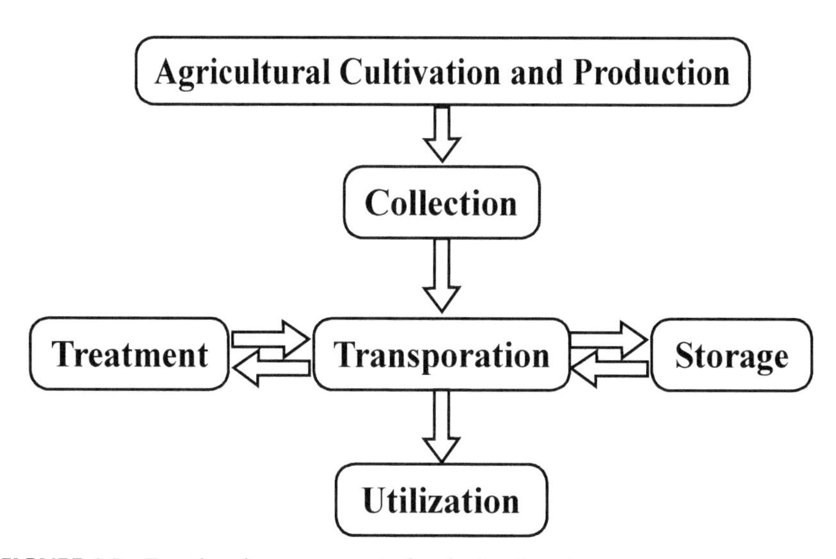

FIGURE 4.2 Functions in management of agricultural wastes.

The COPRAS (COmplex PRoportional ASsessment) method has been widely used when the optimal alternative need to be selected by decision-makers among a pool of alternatives in view of a set of evaluation criteria. In the case of classical COPRAS method for the evaluation process, the weights of criterion along with the ratings of alternatives are precisely known, and the crisp values are employed. Conversely, to model

TABLE 4.1 Selection of Alternatives

Alternatives	Descriptions
Production (A_1)	The production of un-necessary wastes needs to be kept to a minimum level.
Collection (A_2)	The agricultural waste management plan should make-out the collection methods, collection points' locations, collection scheduling, labor requirements, essential structural facilities or equipment, installation, and management costs of the components, and impacts of collection on wastes' consistency.
Transfer (A_3)	The wastes need to be transferred as solids, liquids, or slurry. The agricultural waste management plan must include the consistency analysis of wastes to be moved, transportation methods, distances among transfer points, equipment requirements, and management and installation costs of transfer systems.
Storage (A_4)	It is required to identify the storage period of wastes, storage type and volume, location, storage facility installation costs, management cost of storage processes, and the storage impacts on the consistency of wastes.
Treatment (A_5)	The analysis of the characteristics of the wastes should be included before treatment in the waste management, including the selection of the type, size estimated, location, cost of installation of the treatment facilities, and cost of management in the treatment process.
Utilization (A_6)	Through land application, it is a common practice for recycling the nutrients in the wastes.

real-life decision problems under several conditions, the crisp data happen to be in-capable, and it becomes difficult for the evaluators for precise ratings of alternatives as well as to assign exact weights to the evaluation criterion. Thus, the benefit of using a fuzzy-approach is to find out the relative importance of attributes by the use of fuzzy numbers rather than the numbers precisely. Parezanović et al. (2016) have demonstrated the Fuzzy-COPRAS approach in order to make decisions on mobility measures by evaluating 26 measures. Bekar et al. (2016) have used the COPRAS of alternatives with Grey relations (COPRAS-G) and the Fuzzy-COPRAS method for evaluating the performance measures in total productive maintenance (TPM) strategy. Zarbakhshnia et al. (2018) have proposed a MADM model for ranking and selecting of third-party reverse logistics providers (3PRLPs) by the use of fuzzy step-wise weight assessment ratio analysis (SWARA) for assigning weights to the evaluation criteria. Further, Fuzzy-COPRAS method was used for ranking and selecting the sustainable 3PRLPs in the existence of risk factors. Therefore, to deal with the deficiency in the COPRAS method, the Fuzzy-COPRAS method was used in this study. The Fuzzy-COPRAS assigned the criteria weights and the linguistic-terms represented by fuzzy-numbers were used for evaluating the ratings of alternatives. The steps involved in the Fuzzy-COPRAS method included the following:

- **Step 1:** Defining the linguistic terms.
- **Step 2:** Constructing the fuzzy decision matrix.
- **Step 3:** Determining the weights of criteria.
- **Step 4:** Determining the aggregate fuzzy-rating $\hat{x}_{ij}$ of alternative $A_{i,}$ i = 1, 2,…m under criterion C_j, j = 1, 2,…n.

$$\begin{matrix} & C_1 \ldots\ldots C_2 \ldots\ldots\ldots\ldots C_n \end{matrix}$$

$$\hat{D} = \begin{matrix} A_1 \\ A_2 \\ \vdots \\ A_M \end{matrix} \begin{bmatrix} \tilde{x}_{11} & \tilde{x}_{12} & \cdots & \tilde{x}_{1n} \\ \tilde{x}_{21} & \tilde{x}_{22} & \vdots & \tilde{x}_{2n} \\ \vdots & \vdots & \ddots & \vdots \\ \tilde{x}_{m1} & \tilde{x}_{m2} & \vdots & \tilde{x}_{mn} \end{bmatrix} = , i = 1,2,....,m; j = 1,2,...n \qquad (1)$$

$$\tilde{x}_{ij} = (x_{ij1,} x_{ij2,} x_{ij3});$$

$$x_{ij1,} = \min_k \{x_{ijk1}\}, \; x_{ij2} = \frac{1}{2}\sum_{k=1}^{k} x_{ijk2,} \; x_{ij1,} = \min_k \{x_{ijk3}\}, \qquad (2)$$

where; $\hat{x}_{ijk}$ is the rating of alternative A_i with respect to criterion C_j evaluated by k[th] expert (here k = 8), $\tilde{x}_{ijk} = (\tilde{x}_{ijk1}, \tilde{x}_{ijk2}, \tilde{x}_{ijk3})$.

- **Step 5:** Defuzzifying the obtained aggregated fuzzy decision-matrix and deriving their crisp values. In this analysis for transforming the fuzzy-weights into the crisps-weights applied the center of area method for the calculation of the "best non-fuzzy performance (BPN)" value of the fuzzy-weight of each dimension. The BPN values of the fuzzy-number $\tilde{x}_{ij}$ can be found using the following Eqn. (3).

$$\tilde{x}_{ij} = \tilde{x}_{ij}\,\tilde{x}_{ij} = \frac{[(U\tilde{x}_{ij} - L\tilde{x}_{ij}) + (M\tilde{x}_{ij} - L\tilde{x}_{ij})]}{3} + L\tilde{x}_{ij} \tag{3}$$

- **Step 6:** Normalizing the decision-matrix (f_{ij}). The normalization of the decision-matrix is calculated by dividing each-entry by the largest-entry in each column for eliminating anomalies with different measurement-units, so that all the criteria become dimensionless.
- **Step 7:** Calculating the weighted normalized decision-matrix ($\hat{x}_{ij}$). The fuzzy-weighted normalized-values are calculated by multiplying the weight of evaluation indicator (w_j) with normalized decision-matrices:

$$\hat{x}_{ij} = f_{ij} \cdot w_j \tag{4}$$

- **Step 8:** Summing up of attributed values (P_i) with larger values more preferably (optimization direction is maximization) calculated for each alternative (line of the decision-making matrix):

$$P_i = \sum_{j=1}^{k} \hat{x}_{ij} \tag{5}$$

- **Step 9:** Summing up of attributed-values (R_i) with smaller values more preferably (optimization direction is minimization) calculated for each alternative:

$$R_i = \sum_{j=k+1}^{m} \hat{x}_{ij} \tag{6}$$

where; (m-k) is a number of attributes that need to be minimized.

- **Step 10:** Determining the minimal-value of R_i:

$$R\min = \max_{i} R_i;\ i = 1, \ldots\ldots, n. \tag{7}$$

➢ **Steps 11:** Calculating the relative-weight of each alternative (Q_i):

$$Q_i = P_i + \frac{Rmin\sum_{i=1}^{n} Ri}{Ri\sum_{i=1}^{n}\frac{Rmin}{Ri}} \tag{8}$$

Formula (8) can to be written as:

$$Q_i = P_i + \frac{\sum_{i=1}^{n} Ri}{Ri\sum_{i=1}^{n}\frac{1}{Ri}} \tag{9}$$

➢ **Steps 12:** Determining the optimality-criterion (k):

$$K = \max_{i} Q_i;\ i = \overline{1,n}. \tag{10}$$

➢ **Step 13:** Assigning the priority of the alternatives. The greater the weight (relative weight of alternative) Q_i, the higher is the priority (rank) of the alternatives. The satisfaction degree is high in case of Q_{max}.

➢ **Steps 14:** Calculating the utility degree of each alternative as:

$$N_i = \frac{Qi}{Qmax} \times 100 \tag{11}$$

where; Q_i, and Qmax are the weights obtained from Eqn. (8).

4.3 RESULTS AND DISCUSSION

Every year India produces approximately 61.63 Mt of vegetables and fruits wastes (Table 4.2). It was observed that Maharashtra state in India produces the highest amount of vegetables as well as fruits wastes, followed by Gujarat state. The vegetable wastes are mainly contributed by potato, tomato, okra, brinjal, cabbage onion, and cauliflowers, whereas banana, mango, citrus, papaya, guava, grapes, sapota, and pomegranates are mainly contributing the fruit wastes.

Most of the plant nutrients are generated through crop residues and the recycling of these crop residues into the soil provides beneficial effects on soil fertility as well as improvement in productivity (Table 4.3). Consequently by removing and burning of the crop residues continuously can result in net nutrients losses leading to a high nutrient input cost in the short term and lessening in soil quality as well as productivity in the long term. It has been revealed that on an average 70 to 80% of applied K and 30 to 35% of P & N are built up in the food crops' residues. Moreover, about 80 to 85% of K, 40 to 50% of S, 30 to 35% of P, and 40% of N that is up-taken by rice remains in the vegetative parts at maturity levels (Dobermann and Fairhurst, 2002). Similarly, about 70 to 75% of K, 25 to 30% of P & N, and 35 to 40% of S up-taken by wheat is retained as residues (Singh and Sidhu, 2014). As about 40% of the total dry biomass is constituted by C, so the primary source of organic matters are crop residues which are considered as requisites for sustainability in agricultural ecosystems (Chatterjee et al., 2017). However, the conditions of soils, management of crops, season as well as variety determine the concentration of nutrients in crop residues.

TABLE 4.2 Agricultural Wastes Generation in Different States in India (2010–2011)

State in India	Vegetables Wastes (Mt/year)	Fruits Wastes (Mt/Year)
Madhya Pradesh	2.65	3.71
Maharashtra	15.97	17.17
Gujarat	9.05	7.98
Rajasthan	0.64	0.39
Chhattisgarh	3.14	0.90

Source: Bhattacharjya et al. (2019).

TABLE 4.3 Agricultural Crop Residues' Nutrient Potentials in India (%)

Type of Crop	N	K_2O	P_2O_5	Total
Rice	0.61	1.38	0.18	2.17
Wheat	0.48	1.18	0.16	1.82
Maize	0.52	1.35	0.18	2.05
Sugarcane	0.40	1.28	0.18	1.86
Oilseeds	0.80	0.93	0.21	1.94
Groundnut	1.60	1.37	0.23	3.20
Potato tuber	0.52	1.06	0.21	1.79
Pulses	1.29	1.64	0.36	3.29

Source: Bhattacharjya et al. (2019).

The physical characteristics of agricultural wastes are as shown in Table 4.4, which revealed that the percentage by wet-weight of vegetable wastes (15.6%) is more followed by banana wastes (12.5%), sugarcanes trashes (10.7%), grasses (10.2%) and others (Figure 4.3).

TABLE 4.4 The Physical Characteristics of Agricultural Wastes

Parameters	Percentage by Wet Weight
Dry-leaves	9.8
Vegetable wastes	15.6
Weeds	5.9
Grasses	10.2
Fruits wastes	7.2
Flowers	3.6
Banana wastes	12.5
Sugarcanes trashes	10.7
Coconut wastes	4.3
Eucalyptus	2.2
Parthenium	1.1
Ashes	3
Others	13.8

Source: Lokeshwari and Swamy (2010).

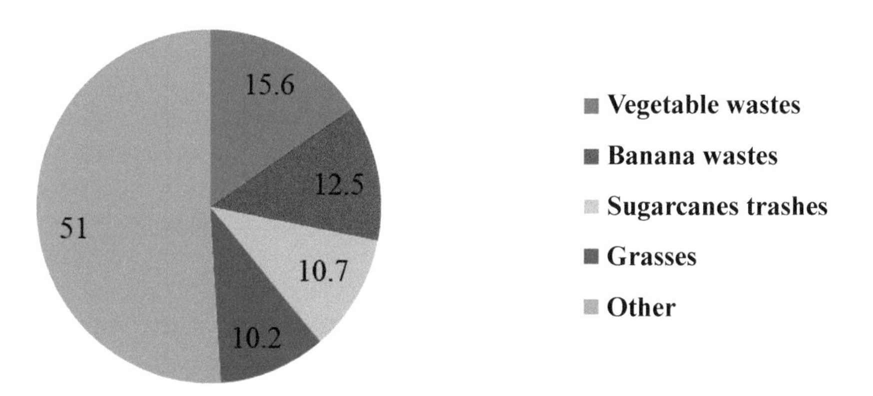

FIGURE 4.3 Percentage by wet-weights of agricultural wastes.

An average of 500 million tons (Mt) of crop residues is generated per year in India, whose majority is used as fodders, fuel for other industrial as well as domestic purposes. However, from surplus crop residues of 140 Mt, about 92 Mt has been burnt every year (NPMCR, 2019). As illustrated in Table 4.5, it can be seen that the generated agricultural wastes in India is more with 500 Mt/Year followed by other selected nations such as Bangladesh (72 Mt/Year), Indonesia (55 Mt/Year), and Myanmar (19 Mt/Year), respectively (NPMCR, 2019; Jeff et al., 2017).

TABLE 4.5 Generated Agricultural Wastes in India and Other Selected Nations in the Same Region

Country	Generated Agricultural Wastes (Million tons/Year)
India	500
Bangladesh	72
Myanmar	19
Indonesia	55

Source: NPMCR (2019); Jeff et al. (2017).

The crop residues created by major crops are as summarized in Table 4.6 that as field residues act as a natural resource contributing traditionally for soil stability as well as fertility through composting, or by directly ploughing into the soil. The irrigation efficiency and control in erosion can be achieved through good management of field residues. Most of the developing countries in Asia have a common practice of burning surplus crop residues (Gadde et al., 2000; Mendoza and Mendoza, 2016). Though burning of residues produces environmental issues, but newer as well as expensive technical assistances are necessary to plow field residues into a larger ground of millions of hectares in a short time duration. The wastes generated from the agri-based industry can be suitably utilized in a variety of industrial processing as well as agro-based applications. On the other hand, the collecting, processing in addition to the transporting costs can be much more than the revenues by the advantageous use of such wastes. Due to the organic composition of agricultural wastes, these crop residues can be used for the societal benefits. India is very prone to high adverse environmental impacts because of the volume of crop residues with un-sustainable management practices (Ross, 2018). Moreover, India has been considered as the second largest country that produces more rice and wheat in the world, resulting in larger volumes of crop residues.

TABLE 4.6 Residues Due to Major Crops

Crop Sources	Compositions
Rice	Husks, and bran
Wheat	Bran, and straws
Maize	Stover, skins, and husks
Sugarcane	Sugarcane tops, bagasse, and molasses
Millet	Stover

Source: Phonbumrung and Khemsawas (1998); Arvanitoyannis (2008).

4.3.1 OUTPUT OF FUZZY-COPRAS METHOD

The criteria names, types, and corresponding weights were as illustrated in Table 4.7.

TABLE 4.7 Different Criteria Weights

Name of Criteria	Type of Criteria	Weights
Economical impacts (C_1)	Minimization	0.77
		0.83
		0.95
Social impacts (C_2)	Minimization	0.62
		0.78
		0.80
Environmental impacts (C_3)	Minimization	0.47
		0.64
		0.78

The initial matrix comprising of the weights of all the criterion and corresponding fuzzy numbers of each alternatives with respect to the criteria were as shown in Table 4.8. Subsequently, the sum of squares and square root (SQRT) values were obtained as illustrated in Table 4.9.

TABLE 4.8 Initial Matrix

	W_1			W_2			W_3		
	0.77	0.83	0.95	0.62	0.78	0.8	0.47	0.64	0.78
	C_1			C_2			C_3		
A_1	50	62.5	75	75	80	85	70	75	80
A_2	55	67.5	80	52	54	56	62	69	76
A_3	60	69	78	75	80	85	70	75	80
A_4	70	81.5	93	54	58	62	55	63.5	72
A_5	84	86.5	89	80	85	90	75	77.5	80
A_6	84	86.5	89	75	80	85	70	75	80

TABLE 4.9 Sum of Squares and SQRT Values

Sum of Squares	**28137**	**34830.3**	**42600**	**28895**	**32705**	**36755**	**27194**	**31674.5**	**36560**
SQRT	324.91			313.62			308.92		
	324.91			313.62			308.92		
	324.91			313.62			308.92		

The normalized matrix and the weighted normalized matrix were as shown in Tables 4.10 and 4.11, respectively.

TABLE 4.10 Normalized Matrix

	W_1			W_2			W_3		
	0.77	0.83	0.95	0.62	0.78	0.8	0.47	0.64	0.78
	C_1			C_2			C_3		
A_1	0.1539	0.1924	0.2308	0.2391	0.2551	0.2710	0.2266	0.2428	0.2590
A_2	0.1693	0.2077	0.2462	0.1658	0.1722	0.1786	0.2007	0.2234	0.2460
A_3	0.1847	0.2124	0.2401	0.2391	0.2551	0.2710	0.2266	0.2428	0.2590
A_4	0.2154	0.2508	0.2862	0.1722	0.1849	0.1977	0.1780	0.2056	0.2331
A_5	0.2585	0.2662	0.2739	0.2551	0.2710	0.2870	0.2428	0.2509	0.2590
A_6	0.2585	0.2662	0.2739	0.2391	0.2551	0.2710	0.2266	0.2428	0.2590

The ranking of alternatives were based on the higher values of N_i as summarized in Table 4.12, and it was observed that the agricultural waste management function such as "Storage (A_4)" ranked first requiring more attention, followed by Collection (A_2), Utilization (A_6), Treatment (A_5), Transfer (A_3), and Production (A_1), respectively (Figure 4.4).

TABLE 4.11 Weighted Normalized Matrix

	C_1			C_2			C_3		
A_1	0.1185	0.1597	0.2193	0.1483	0.1990	0.2168	0.1065	0.1554	0.2020
A_2	0.1303	0.1724	0.2339	0.1028	0.1343	0.1428	0.0943	0.1430	0.1919
A_3	0.1422	0.1763	0.2281	0.1483	0.1990	0.2168	0.1065	0.1554	0.2020
A_4	0.1659	0.2082	0.2719	0.1068	0.1443	0.1582	0.0837	0.1316	0.1818
A_5	0.1991	0.2210	0.2602	0.1582	0.2114	0.2296	0.1141	0.1606	0.2020
A_6	0.1991	0.2210	0.2602	0.1483	0.1990	0.2168	0.1065	0.1554	0.2020

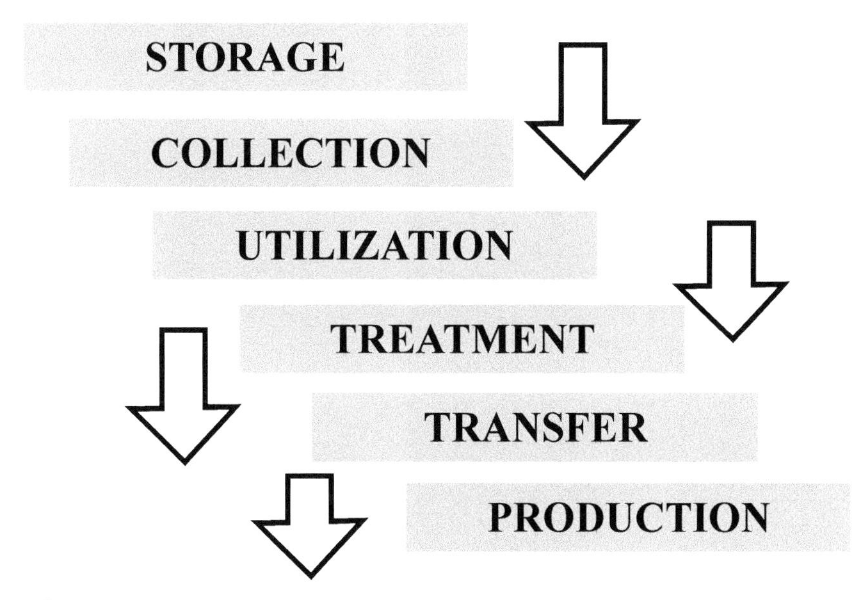

FIGURE 4.4 Functions in management of agricultural wastes.

Moreover, in order to control the excessive food wastes from agricultural sources, it is highly desirable to consider adequate precautions and control over its production as well as consumption (Figure 4.5).

TABLE 4.12 Ranking of Alternatives

Alternatives	P_j			R_j			Q_j			Non-Fuzzy Q_j	N_j	Ranking
A_1	0.2220	0.3689	0.4494	0.2548	0.3544	0.4188	0.296	0.674	1.524	0.831	78.79	6
A_2	0.2269	0.3682	0.4497	0.1971	0.2773	0.3347	0.319	0.758	1.838	0.972	92.12	2
A_3	0.2388	0.3788	0.4582	0.2548	0.3544	0.4188	0.313	0.684	1.533	0.843	79.90	5
A_4	0.2763	0.4377	0.5308	0.1904	0.2758	0.3400	0.367	0.830	1.968	1.055	100	1
A_5	0.3025	0.4370	0.5047	0.2723	0.3720	0.4316	0.374	0.728	1.510	0.871	82.52	4
A_6	0.3163	0.4640	0.5335	0.2548	0.3544	0.4188	0.390	0.769	1.608	0.922	87.43	3

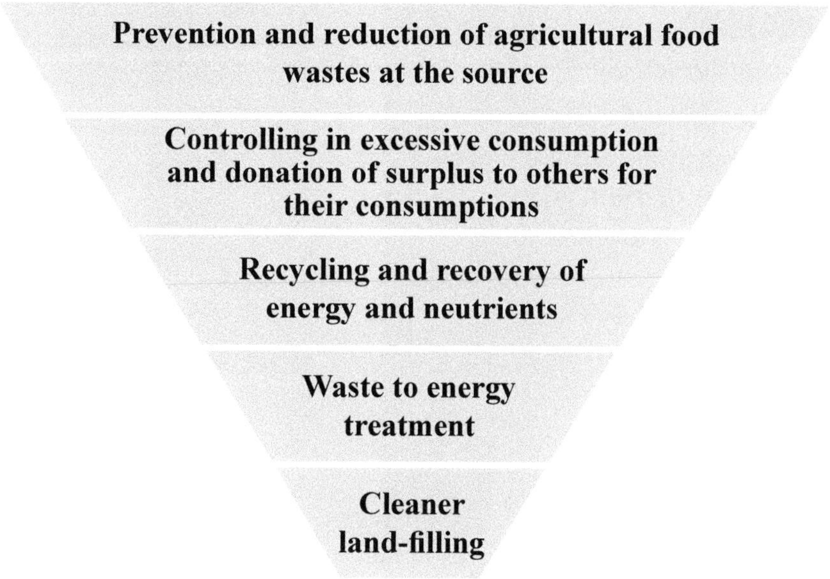

Prevention and reduction of agricultural food wastes at the source

Controlling in excessive consumption and donation of surplus to others for their consumptions

Recycling and recovery of energy and neutrients

Waste to energy treatment

Cleaner land-filling

FIGURE 4.5 Effective food control for sustainable agriculture.

4.4 CONCLUSION

The farming divisions in India contribute around 17% of the nation's GDP and gives work to roughly two-thirds of the populace. However, due to the under-development of the food sectors, its potential has not been tapped in India. There are different issues to be managed in using the opportunities for the agri-business industry. The Indian products are confronting extreme challenges in the universal market with regards to quality, pesticide deposits, assortments with more timeframe of realistic usability, packaging, and so forth. This infers Indian produce should be progressively focused to confront the global challenge, which again requests the upkeep of value benchmarks all through the value chain, including great rural practices. However, it has been discovered a major amount of fruits, vegetables, and crops are wasted due to un-awareness as well as an effective agricultural waste management plan, so the findings and suggestions in this study will enable the decision-makers to take appropriate steps in this regard.

KEYWORDS

- **agricultural waste management**
- **analytic hierarchical process**
- **fuzzy-COPRAS**
- **life-cycle-assessment**
- **multicriteria decision-making**
- **municipal solid waste**

REFERENCES

Agamuthu, P., (2009). *Challenges and Opportunities in Agro-Waste Management: An Asian Perspective*. Inaugural meeting of First Regional 3R Forum in Asia. Tokyo, Japan.

Alkaradaghi, K., Ali, S. S., Al-Ansari, N., Laue, J., & Chabuk, A., (2019). Landfill Site Selection Using MCDM Methods and GIS in the Sulaimaniyah Governorate, Iraq. *Sustainability, 11,* 4530. doi: 10.3390/su11174530.

Alzate, C. A., Sa'nchez, O. J., Ramı'rez, J. A., & Rinco'n, L. E., (2004). Biodegradation of organic waste from market places. *Rev. Colomb. Biotecnol., 6,* 78–89. Available at: https://revistas.unal.edu.co/index.php/biotecnologia/article/view/529 (accessed on 15 March 2021).

Amoding, A., (2007). *Supply Potential and Agronomic Value of Urban Market Crop Wastes*. PhD Thesis, Makerere University.

Arvanitoyannis, I. S., & Tserkezou, P., (2008). Wheat, barley and oat waste: A comparative and critical presentation of methods and potential uses of treated waste. *Int. J. Food Sci. Technol., 43,* 694–725.

Batarseh, E. S., Reinhart, D. R., & Berge, N. D., (2006). Sustainable disposal of municipal solid waste: Post bioreactor landfill polishing. *Waste Management, 20,* 2170–2176.

Bekar, E. T., Cakmakci, M., & Kahraman, C., (2016). Fuzzy COPRAS method for performance measurement in total productive maintenance: A comparative analysis. *Journal of Business Economics and Management, 17*(5), 663–684. doi: 10.3846/16111699.2016.1202314.

Bhattacharjya, S., Sahu, A., Manna, M. C., & Patra, A. K., (2019). Potential of surplus crop residues, horticultural waste and animal excreta as a nutrient source in the central and western regions of India. *Current Science, 116*(8), 1314–1323. doi: 10.18520/cs/v116/i8/1314-1323.

Bhuvaneshwari, S., Hettiarachchi, H., & Meegoda, J. N., (2019). Crop residue burning in India: Policy challenges and potential solutions. *Int. J. Environ. Res. Public Health, 16,* 832. doi: 10.3390/ijerph16050832.

Brouwer, J., & Powell, J. M., (1998). Increasing nutrient use efficiency in West African agriculture: The impact of micro-topography on nutrient leaching from cattle and sheep manure. *Agriculture, Ecosystems and Environment, 71,* 229–239.

Burns, J. C., King, L. D., & Westerman, P. W., (1990). Long term swine lagoon effluent applications on to coastal bermudagrass: I. Yield, quality, and elemental removal. *Journal of Environmental Quality, 19,* 749–756.

Büyükbektaş, F., & Varınca, K. B., (2010). *Integrated Waste Management Concept and Waste Framework Directive* (pp. 1–6). EU Harmonization Process.

CAST, (1975). *Ruminants as Food Producers: Now and for the Future* (Vol. 4, pp. 1–13). Council for Agricultural Science and Technology. Special Publication.

Chatterjee, R., Gajjela, S., & Thirumdasu, R. K., (2017). Recycling of organic wastes for sustainable soil health and crop growth. *Int. J. Waste Resour., 7,* 296.

Coelho, L. M. G., Lange, L. C., & Coelho, H. M. G., (2017). Multi-criteria decision making to support waste management: A critical review of current practices and methods. *Waste Management and Research, 35*(1), 3–28. doi: 10.1177/0734242X16664024.

Debnath, A., Roy, J., Kar, S., Zavadskas, E. K., & Antucheviciene, J., (2017). A hybrid MCDM approach for strategic project portfolio selection of agro by-products. *Sustainability, 9,* 1302. doi: 10.3390/su9081302.

Department of Environment, (2010). *National 3R Strategy for Waste Management.* Ministry of Environment and Forests, Government of the People's Republic of Bangladesh.

Dien, B. V., & Vong, V. D., (2006). *Analysis of Pesticide Compound Residues in Some Water Sources in the Province of Gia Lai and Dak Lak.* Vietnam Food Administrator.

Dobermann, A., & Fairhurst, T. H., (2002). Rice straw management. *Better Crops Int., 16,* 7–9.

Ekmekçioğlu, M., Kaya, T., & Kahraman, C., (2010). Fuzzy multicriteria disposal method and site selection for municipal solid waste. *Waste Management, 30,* 1729–1736.

Eskandari, M., Homaee, M., & Mahmodi, S., (2012). An integrated multicriteria approach for landfill siting in a conflicting environmental, economic and socio-cultural area. *Waste Management, 32,* 1528–1538.

Ezcurra, A. I., Ortiz, D. Z., Pham, V. D., & Lacaux, J. P., (2001). Cereal waste burning pollution observed in the town of Vitoria (northern Spain). *Atmospheric Environment, 35,* 1377–1386.

Gadde, B., Bonnet, S., Menke, C., & Garivait, S., (2000). Air pollutant emissions from rice straw open field burning in India, Thailand and the Philippines. *Environ. Pollut., 157,* 1554–1558.

Hai, H. T., & Tuyet, N. T. A., (2010). *Benefits of the 3R Approach for Agricultural Waste Management (AWM) in Vietnam.* Under the framework of joint project on Asia resource circulation policy research working paper series, Institute for Global Environmental Strategies supported by the Ministry of Environment, Japan.

Hanan, D., Burnley, S., & Cooke, D., (2012). A multicriteria decision analysis assessment of waste paper management options. *Waste Management,* 1–8.

Hargreaves, J. C., Adl, M. S., & Warman, P. R., (2008). A review of the use of composted municipal solid waste in agriculture. *Agriculture, Ecosystems and Environment, 123*(1–3), 1–14.

Hegg, D. A., Radke, L. F., Hobbs, P. V., Brock, C. A., & Riggan, P. J., (1987). Nitrogen and sulfur emissions from the burning of forest products near large urban areas. *Journal of Geophysical Research, 92,* 14701–14709.

Jakobsen, S., (1995). Aerobic decomposition of organic wastes 2. Value of compost as fertilizer. *Resources, Conservation and Recycling, 13,* 57–71.

Jeetah, P., Golaup, N., & Buddynauth, K., (2015). Production of cardboard from waste rice husk. *J. Environ. Chem. Eng., 3*(1), 52–59.

Jeff, S., Prasad, M., & Agamuthu, P., (2017). Asia waste management outlook. *UNEP Asian Waste Management Outlook.* United Nations Environment Program: Nairobi, Kenya.

Karmperis, A. C., Aravossis, K., Tatsiopoulos, I. P., et al., (2013). Decision support models for solid waste management: Review and game-theoretic approaches. *Waste Management, 33,* 1290–1301.

Karmperis, A. C., Sotirchos, A., Aravossis, K., & Tatsiopoulos, I. P., (2012). Waste management project's alternatives: A risk-based multicriteria assessment (RBMCA) approach. *Waste Management, 32,* 194–212.

Katongole, C. B., (2009). *Developing Rations for Meat Goats Based on Some Urban Market Crop Wastes.* PhD Thesis, Makerere University, Uganda.

Khan, S., & Faisal, M. N., (2008). An analytic network process model for municipal solid waste disposal options. *Waste Management, 28,* 1500–1508.

Kharat, M. G., Kamble, S. S., Kamble, S. J., & Dhume, S. M., (2016). Modeling landfill site selection using an integrated fuzzy MCDM approach. *Model. Earth Syst. Environ., 2,* 53. doi: 10.1007/s40808-016-0106-x.

Kumar, S., Smith, S. R., Fowler, G., Velis, C., Kumar, S. J., Arya, S. R., Kumar, R., & Cheeseman, C., (2017). Challenges and opportunities associated with waste management in India. *R. Soc. Open Sci., 4,* 160764. Available at: http://dx.doi.org/10.1098/rsos.160764.

Kumar, U., & Bandyopadhyay, M., (2006). Sorption of cadmium from aqueous solution using pretreated rice husk. *Bioresour. Technol., 97*(1), 104–109.

Lacaux, J. P., Loemba-Ndembi, J., Lefeivre, B., Cros, B., & Delmas, R., (1992). Biogenic emissions and biomass burning influences on the chemistry of the fog water and stratiform precipitations in the African equatorial forest. *Atmospheric Environment, 26*(A/4), 541–551.

Lennartsson, P. R., Erlandsson, P., & Taherzadeh, M. J., (2014). Integration of the first and second-generation bioethanol processes and the importance of by-products. *Bioresour. Technol., 165,* 3–8.

Lim, T., Heber, A. J., Ni, J., Sutton, A. L., & Shao, P., (2003). Odor and gas release from anaerobic treatment lagoons for swine manure. *J. Environ. Qual., 32*(2), 406–416.

Lokeshwari, M., & Swamy, C. N., (2010). Waste to wealth agriculture solid waste management study. *Poll Res., 29*(3), 129–133.

Lowe, P. D., (1995). Social issues and animal wastes: A European perspective. In: *Proceedings of International Livestock Odor Conference* (pp. 168–171). Iowa State University College of Agriculture, America.

Mackie, R. I., Stroot, P. G., & Varel, V. H., (1998). Biochemical identification and biological origin of key odor components in livestock waste. *Journal of Animal. Science, 76,* 1331–1342.

Mardani, A., Jusoh, A., Nor, K. M. D., Khalifah, Z., Zakwan, N., & Valipour, A., (2015). Multiple criteria decision-making techniques and their applications: A review of the literature from 2000 to 2014. *Economic Research-Ekonomska Istraživanja, 28*(1), 516–571. doi: 10.1080/1331677X.2015.1075139.

Mendoza, T. C., & Mendoza, B. C., (2016). A review of sustainability challenges of biomass for energy, focus in the Philippines. *Agric. Technol., 12,* 281–310.

Menikpura, S. N. M., Sang-Arun, J., & Bengtson, M., (2013). Integrated solid waste management: An approach for enhancing climate co-benefits through resource recovery. *Journal of Cleaner Production,* 1–9.

Milutinović, B., Stefanović, G., Dassisti, M., et al., (2014). Multicriteria analysis as a tool for sustainability assessment of a waste management model. *Energy, 74,* 190–201.

Moncada, J., Tamayo, J. A., & Cardona, C. A., (2014). Integrating first, second, and third-generation biorefineries: Incorporating microalgae into the sugarcane biorefinery. *Chem. Eng. Sci., 118,* 126–140.

Morrissey, A. J., & Browne, J., (2004). Waste management models and their application to sustainable waste management. *Waste Management, 24,* 297–308.

Mukhtar, S., Borhan, M. S., Rahman, S., & Zhu, J., (2010). Evaluation of a field-scale surface aeration system in an anaerobic poultry lagoon. *Appl. Eng. Agric., 26,* 307–318.

Muşdal, H. (2007). Tıbbi atıkları işleme ve bertaraf etme teknolojisi seçme problemine bulanık analitik hiyerarşi prosesi ve bulanık analitik ağ prosesi yaklaşımı. MSc Thesis (in Turkish), Yıldız Technical University, 2007.

NPMCR, (2019). Available online: https://agricoop.nic.in/sites/default/files/NPMCR_1.pdf (accessed on 8 October 2019).

Obi, F. O., Ugwuishiwu, B. O., & Nwakaire, J. N., (2016). Agricultural waste concept, generation, utilization and management. *Nigerian Journal of Technology, 35*(4),957–964.

Ohman, K. V. H., Hettiaratchi, J. P. A., Ruwanpura, J., Balakrishnan, J., & Achari, G., (2007). Development of a landfill model to prioritize design and operating objectives. *Environmental Monitoring and Assessment, 135,* 85–97.

Oltjen, J. W., & Beckett, J. L., (1996). Role of ruminant livestock in sustainable agricultural systems. *Journal of Animal Science, 74,* 1406–1409.

Parezanović, T., Bojković, N., Petrović, M., & Tarle, S. P., (2016). Evaluation of sustainable mobility measures using fuzzy COPRAS method. *Management, 78,* 53–62.

Phonbumrung, T., & Khemsawas, C., (1998). Agricultural crop residue. In: *Proceedings of the Sixth Meeting of Regional Working Group on Grazing and Feed Resources for Southeast Asia* (pp. 183–187). Legaspi, Philippines.

Pinzi, S., & Dorado, M. P., (2012). Feedstocks for advanced biodiesel production. *Advances in Biodiesel Production* (pp. 204–231). Woodhead Publishing Limited, Cambridge.

Pires, A., Chang, N., & Martinho, G., (2011). Solid waste management in European countries: A review of systems analysis techniques. *Journal of Environmental Management, 92,* 1033–1050.

Prada, A., & Corte's, C. E., (2010). Thermal decomposition of rice husk: An alternative integral use. *Rev. Orinoquia, 3*(1), 155–170.

Quiceno, V. D., & Mosquera, G. M. Y., (2010). Alternativas Tecnologicas para el uso de la cascarilla de arroz como combustible. Universidad Auto'noma de Occidente.

Rincon, L. E., Moncada, B., J., & Cardona, A. C. A., (2013). *Catalytic Systems for Integral Transformations of Oil Plants Through Biorefinery Concept* (1st edn.). Universidad Nacional de Colombia Sede Manizales, Manizales.

Ross, S., (2018). Countries that produce the most food. *Investopedia.* Available online: https://www.investopedia.com/articles/investing/100615/4-countries-produce-most-food.asp#ixzz5WRqV85mY (accessed on 16 November 2019).

Sabiiti, E. N., Bareeba, F., Sporndly, E., Tenywa, J. S., Ledin, S., Ottabong, E., Kyamanywa, S., et al., (2005). Urban market garbage: A resource for sustainable crop/

livestock production system and the environment in Uganda. *A Paper Presented at the International Conference, Wastes-The Social Context*. Edmonton, Canada.

Sabiiti, E. N., Bareeba, F., Sporndly, E., Tenywa, J. S., Ledin, S., Ottabong, E., Kyamanywa, S., et al., (2004). Urban market garbage: A hidden resource for sustainable urban/peri-urban agriculture and the environment in Uganda. *The Uganda Journal, 50,* 102–109.

Sadh, P. K., Duhan, S., & Duhan, J. S., (2018). Agro-industrial wastes and their utilization using solid-state fermentation: A review. *Bioresources and Bioprocessing, 5,* 1. https://doi.org/10.1186/s40643-017-0187-z.

Schulte, D. D., (1997). Critical parameters for emissions. In: Voermans, J. A. M., & Monteny, G. J., (eds.), *Proceedings of Ammonia and Odor Emissions from Animal Production Facilities*. NVTL Publishing, Rosmalen, The Netherlands.

Si, S. L., You, X. Y., Liu, H. C., & Zhang, P., (2018). DEMATEL technique: A systematic review of the state-of-the-art literature on methodologies and applications. *Mathematical Problems in Engineering,* Article ID: 3696457. Available at: https://doi.org/10.1155/2018/3696457.

Singh, Y., & Sidhu, H. S., (2014). Management of cereal crop residues for sustainable rice-wheat production system in the Indo-Gangetic plains of India. *Proc Indian Natl. Sci. Acad., 80,* 95–114.

Soltani, A., Hewage, K., Reza, B., et al., (2015). Multiple stakeholders in multicriteria decision-making in the context of municipal solid waste management: A review. *Waste Management, 35,* 318–328.

Soltani, N., Bahrami, A., Pech-Canul, M. I., & Gonza'lez, L. A., (2014). Review on the physicochemical treatments of rice husk for production of advanced materials. *Chem. Eng. J., 264,* 899–935.

Tekel, A., (2007). Katı atık yönetiminde stratejik planlama. *Journal of Contemporary Local Governments, 16*(3), 71–83.

Tezçakar, M., & Can, O., (2011). Atıktan enerji eldesinde termal bertaraf teknolojileri. *Recydia A.Ş.,* 1–6.

Tumuhairwe, J. B., Tenywa, J. S., Otabbong, E., & Ledin, S., (2009). Comparison of four low-technology composting methods for market crop wastes. *Waste Management, 29,* 2274–2281.

USDA, (2012). *Agricultural Waste Management Field Handbook.* United States Department of Agriculture, Soil conservation Service. Accessed from: https://www.nrcs.usda.gov/wps/portal/nrcs/detailfull/national/water/?&cid=stelprdb1045935 (accessed on 18 February 2021).

Westerman, P. W., & Bicudo, J. R., (2005). Management considerations for organic waste use in agriculture. *Bioresource Technology, 96,* 215–221.

Yang, L., Xu, F., Ge, X., & Li, Y., (2015). Challenges and strategies for solid-state anaerobic digestion of lignocellulosic biomass. *Renew Sustain. Energy Rev., 44,* 824–834.

Yay, A. S. E., (2015). Application of life cycle assessment (LCA) for municipal solid waste management: A case study of Sakarya. *Journal of Cleaner Production, 94,* 284–293.

Zarbakhshnia, N., Soleimani, H., & Ghaderi, H., (2018). Sustainable third-party reverse logistics provider evaluation and selection using fuzzy SWARA and developed fuzzy COPRAS in the presence of risk criteria. *Applied Soft Computing, 65,* 307–319.

Zurbrügg, C., Gfrerer, M., Ashadi, H., Brenner, W., & Küper, D., (2012). Determinants of sustainability in solid waste management-the Gianyar waste recovery project in Indonesia. *Waste Management, 32,* 2126–2133.

CHAPTER 5

MAKING E-WASTE MANAGEMENT SUCCESSFUL

ABSTRACT

This study proposed a hybrid MCDM approach based on the Fuzzy-TOPSIS method to evaluate the problems and solutions to execute India's e-waste management strategies. Based on the output, the linking up the activities of the informal sector with the formal sectors ranked first, which was followed by addressing a safe disposal of e-waste (EW) both internal and external to domestic levels, imposing hazardous e-waste disposal fees from manufacturers and consumers, adopting a consultative process for an effective e-waste management plan, promoting the awareness program on recycling and disposal of EW, providing incentives and subsidized schemes to the public or industries associated with recycling and disposal of EW, restriction and domestic legal frameworks on importing of EW, and attracting investments in e-waste management sectors, respectively. Moreover, the proposed Fuzzy-TOPSIS method provides an accurate, systematic, and efficient decision tool that enables policymakers to make sound decisions.

5.1 INTRODUCTION

The modern spectacular developments in recent times have enhanced the quality of our lives without a doubt. But, these have resulted in manifold problems at the same time that includes the problems of hazardous wastes in massive amount and other electric products generated wastes posing greater threats to the human health as well as environment. The issue associated with the proper management of these wastes is critical for the livelihood, health, and environment protection. Therefore, this is not only a serious concern to the modern societies, but also requires coordinated efforts in order to achieve sustainable development. According to Hazardous Wastes (Management and Handling) Amendment Rules (2003) in India, ten states were reported of contributing to 70% of the total generated e-wastes (EW), while more than 60% of the total EW were generated by 65 cities. Among the EW generating states and cities, Maharashtra and Mumbai were reported at first ranks. Thus, an effort has been made in this study to make a selection of the most appropriate strategies for effective e-waste management in India. Mundada et al. (2004) have discussed the environmental and occupational hazards associated with the processing of EW by considering hazardous materials along with their composition, and handling as well as processing methods. Sinha-Khetriwal et al. (2005) have made a comparison of e-waste recycling in India as well as Switzerland, and suggested for more quantitative measures in e-waste recycling areas. Different researchers define E-waste or "Waste of Electrical and Electronic Equipment (WEEE)" which has been a fastest-growing streams of wastes (Cairns, 2005) in various ways. For example, as outdated, end-of-life (EoL), or discarded appliances using electricity (Davis and Heart, 2007; Ewasteguide, 2009), and as "Electronic products which no longer satisfies the requirements of the initial purchasers" (Peralta and Fontanos, 2006). The waste generated depends upon the demographic, socio-economic, and geographical perimeters, etc., influences the waste generations (Beigl et al., 2008). More informed e-waste management as well as policy decisions can be made through the successful establishment of baseline information levels in order to achieve sustainable development goals (SDGs) related to waste management (Manga et al., 2008). Correspondingly, the reduction of e-waste generation and its environmental impacts can be achieved through effective management of EW by establishing separate environmentally friendly collection channels (Babu et al., 2007). By analyzing the e-waste disposal systems in developing countries like China (He et al., 2006), Brazil (Oliveira et al., 2012), and India (Dwivedy and Mittal, 2012), focusing on the difficulties in implementation or enforcement of existing regulations as well as clean technologies were revealed due to lack of awareness and capacity building. In contrast, sophisticated, high-cost, and less hazardous disposal schemes have been devised by developed countries. The regulations that guide the disposition of e-waste in developed countries are stringent having effective monitoring, whereas in developing countries, these are mostly fragmented followed by lack of monitoring (Gullett et al., 2007).

Although there is an availability of many technical solutions to be adopted for e-waste management, but the prerequisite conditions like legislation, manpower, logistics, and collection systems are required to be prepared (Monika and Kishore 2010). Most of the EW has been reported to be collected and recycled in the in-formal sector like many other developing countries (Rajya Sabha, 2011; Wath et al., 2011), which is 95% of generated EW (GTZ-MAIT, 2007). Moreover, serious risks to human health in addition to the environment has been revealed to be due to the hazardous nature of EW and the un-scientific practices used by the unskilled as well as semiskilled workers in the informal sectors (Bandyopadhyay, 2008, 2010; Manomaivibool, 2009; Pradhan and Kumar, 2014). Electronic waste may be defined as abandoned computers, electronic office equipment, electronic entertainment devices, refrigerator, television sets, and mobile phones, etc. Used electronic devices destined for reusing, reselling, salvaging, disposal or recycling is included in this definition. The term "e-waste" has been applied by several public policy advocates to all surplus electronics (Shagun et al., 2013). Borthakur and Sinha (2013) have made an attempt in order to identify the variety of stakeholders' right from the production of electrical and electronic equipment to the final disposal of EW in the "E-waste management system" in India. Bhat and Patil (2014) have studied the e-waste disposal and awareness practices among residents of Pune city, and found the consumer awareness as very good and at the superficial level, but lack of awareness of the collection centers, correct disposal practices, and the E-waste rules. Bhat and Patil (2014) have reported of lack of consumer awareness in Pune city (India) about the collection centers of e-waste, rules, and different appropriate e-waste disposal practices. As per the Frost and Sullivan Report (2015), in India, the cities like Pune as well as Mumbai and Navi Mumbai has generated 12,300 MT and 61,500 MT of EW in 2014 (Frost and Sullivan Report, 2015). Jayapradha (2015) has carried out a study on the health hazards of e-waste in addition to the management methods to handle those hazardous wastes. Murray et al. (2015) have suggested the concept of "circular economy" as a distinct way of consumption, as well as an economic model that acts as a tool to operationalize the sustainable development concept. For the recycling industry and also in order to formulate government policies, some essential fundamental information on e-waste include its' amount generated, resource transfer and composition (Zeng et al., 2016b). However, acquiring such effective information remains a challenge which needs adequate addressing (Tran et al., 2016). Zeng et al. (2017) have proposed three different levels for E-waste management that should be addressed at: mesoscopic (material), microscopic (substance), and macroscopic (product and component), respectively. A significant contribution can be made towards valuable resources conservation and sustainable energy utilization through recycling of EW (Zeng et al., 2016a). Because of the presence of a number of un-authorized sectors, and direct connection with environmental problems, the proper e-waste recycling is very important for the sustainable development of a developing country (Borthakur and Govind, 2017). E-waste contains different heavy metals, including a variety of hazardous substances that causes health and environmental risks if not properly managed (Awasthi et al., 2016a). A serious issue has been revealed as the adverse effects on the labor engaged in informal recycling (Awasthi et al., 2016b; Fowler, 2017). For any effective E-waste management initiative, the E-waste disposal behavior as well as awareness of consumers plays a vital role (Borthakur and Singh, 2016). Additionally, an omnipresent ambivalence exists regarding responsible E-waste disposals among Indian citizens (Borthakur and Singh, 2017). Arya et al. (2018) have conducted a pilot study in northern Indian states such as Haryana, Punjab, Uttar Pradesh, Himachal Pradesh, Chandigarh, and Delhi, to find out the level of awareness among the consumers as well as recycler about the hazardous effect of EW. A more environmentally focused and sound e-waste management in India is required with an active support from all the participants involved in the e-waste flow chain (Awasthi et al., 2018). Bakhiyi et al. (2018) have revealed through review of several challenges associated with e-waste management that included lack of harmonization in definitions of e-waste, the leftover toxic potentials of illicit or restricted harmful elements like heavy metals, and bio-accumulative organic composites, continuation in the growth of e-waste volumes, challenging e-recycling practices, damage of informal e-recycling systems combined with unforeseeable patterns as well as complexity of illicit e-waste trades, and the weakness in the formal e-recycling sectors. Likewise, Tansel (2017) has argued of having major challenges in e-waste management in spite of an increase in the market for recycled materials, such as the need of infrastructure for collecting and separating of EW, need of accounting mechanism for cross-boundary transports, and need of awareness as well as training for both safe handling and processing of materials during its' recovery, respectively. Mishra et al. (2017) have studied the awareness level amongst informal handlers about electronic waste along with its health hazards in Musheerabad, Hyderabad in India. It was observed that out of total 104 considered handlers, about 72% of the handlers were un-known about the electronic waste meaning, 71% were not aware of associated risks with health, and 85% were not using any protective gear. Whereas, 16% acknowledged regarding health issues because of improper handling of EW, and 77% felt appropriateness of their e-waste handling practices. It was observed in a study carried out in Kochi City (India) that there was a lack of awareness among the majority of the respondents about the concerns with used or EoL electronics (Anusree and Balasubramanian, 2019). India's first e-waste regulation of India known as "E-waste (Management and Handling) Rules (2011)" used EPR approach with its requirement for the electronic products producers' for setting up collection centers, but the rules have been largely ineffective to improve the existing practices (Bhaskar and Turaga, 2018). Further, with amendments in 2016, its implementation for 7 years had a limited impact on the e-waste management system in the country (Turaga, 2019). In a case study about e-waste carried out in Nagpur (India), it was reported that with the increase in the use of technologies and latest gadgets, a large amount of electronic wastes will be generated, contributing about 27% of total generated wastes. Therefore, it is a basic requirement for minimizing the generation of wastes and also recycling of recyclable wastes (Raut et al., 2019). Implementation of an economic incentive based on the "electronic bonus card system (EBCS)" has a number of benefits like compensating the transaction costs of proper collection for the consumers and satisfying the consumer perception of having residual value of "electrical and electronic equipment (EoL EEE)." However, the cooperation of various stakeholders including electronics producers,' and national as well as international authorities are required for the application of the EBCS motivation technology (Shevchenko et al., 2019). The discarded electronic un-used materials as EW is as shown in Figure 5.1. Thus, there should be control over the disposal and discard of un-used EW (Figure 5.2).

Some examples of EW are as summarized below:

- Air-conditioners;
- Computers/note-books;
- Digital/other cameras;
- DVD/VCD players;
- Electrical-fans;
- Electronic game-devices;

FIGURE 5.1 E-wastes.

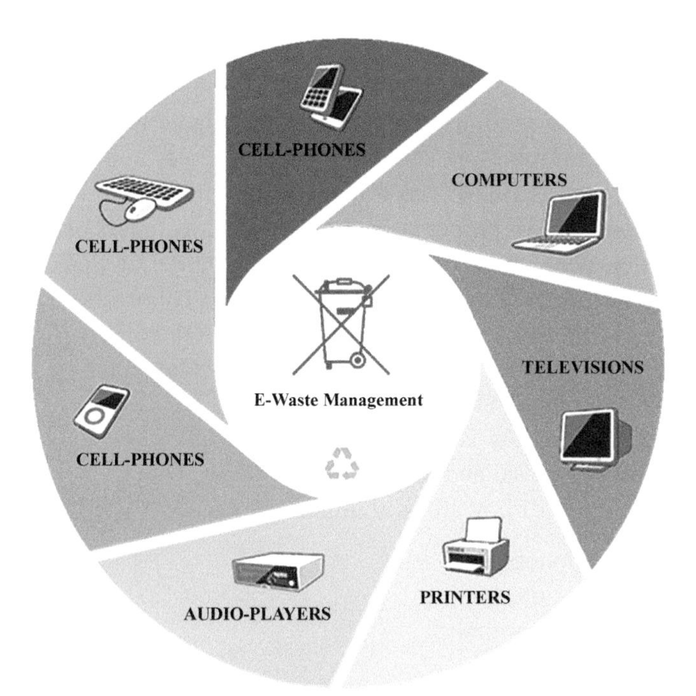

FIGURE 5.2 Un-necessary disposal control of e-wastes.

- Facsimile-machines;
- Irons;
- Oven and microwaves;
- Photostat machines;
- Printers and other accessories;
- Radio/Hi-Fi;
- Refrigerators;
- Rice-cookers;
- Stereo/audio device;
- Telephones/cellphones;
- Televisions;
- Vacuum-cleaners;
- Video-camera;
- Washing-machines.

The weights of different materials present in EW (Chaturvedi, 2016) is as shown in Table 5.1.

5.1.1 INTERNATIONAL PERSPECTIVES ON E-WASTES (EW)

Today, 8% of municipal waste are constituted with the ever increasing "waste electrical and electronic equipment (WEEE)" which has been one of the fastest-growing waste fractions. In recent decades, the use of electronic devices has multiplied, and the quantities of electronic devices, such as mobile phones, PCs, and entertainment-electronics are growing rapidly throughout the world. An approximate amount of 20 million PCs became obsolete by an estimate in 1994, and by 2004, this figure was increased to over 100 million PCs. Further, between 1994 and 2003, about 500 million PCs reached their service lives end. Moreover, it may be noted that about 500 million PCs contain approximately 718,000 t of lead, 287 t of mercury, 2,872,000 t of plastics, and 1363 t of cadmium (Puckett and Smith, 2002). The global market for PCs is far from saturation with a rapid decrement in the average life-span of a PCs (Culver, 2005). An estimation in 2005 revealed of about 130 million mobile-phones to be retired (O'Connell, 2002). The transboundary-movement of electronic wastes has been regulated by the Basel-Convention (UNEP, 1989), and due to the presence of highly toxic substances in EW such as lead, cadmium, and mercury, it is considered to be dangerous to human-being as well as the environment (EU,

TABLE 5.1 Weight of Different Materials in E-Wastes

Appliances	Average Weight in kg	% Weight of Iron (Fe)	% Weight of Glass	% Weight of Plastic	% Weight of Non-Fe-Metal	% Weight of Electronic Components	% Weight of Others
TV sets	36.2	5.3	62	22.9	5.4	0.9	3.5
Cellular telephones	0.08–0.100	8	10.6	59.6	10.6	–	1.6
Personal computers	29.6	53.3	15	23.3	8.4	17.3	0.7
Washing machines	40–47	59.8	2.6	1.5	4.6	–	31.5
Refrigerators and freezers	48	64.4	1.5	13	6	–	15.1

2002b). However, the recovery of some valuable substances such as gold and copper metals from EW has become a profitable-business worldwide that result in international, transboundary trade in EW. Countries such as India and China are facing a rapid increase in the amount of EW from both domestic-generation and illegal-imports. While, it is a source of livelihood for the rural as well as urban poor peoples who are not aware about the potential associated risks, un-known about better practices, and having no access to investment-capital in order to finance profitable-improvements. The waste-electrical and electronic-equipment (WEEE) consists of the ten categories as defined in the "Directive 2002/96/EC of the European Parliament and of the Council (January 2003)" (EU, 2002a), as illustrated in Table 5.2.

TABLE 5.2 Categories of WEEE as Defined in the EU Directive on Waste-Electrical and Electronic-Equipment (EU, 2002a)

Categories	Labels
Automated-dispensers	Dispenser
Large household appliances	Large-HH
Consumers-equipment	CE
Medical-devices exclusive of all infected as well as embedded products	Medical-equipment
Lighting-equipment	Lighting
IT and telecommunications equipment	ICT
Monitoring and control instrument	M & C
Electrical and electronic tools, exclusive of large-scale stationary industrial-tools	E & E tools
Small-household-appliances	Small-HH
Toys, leisure, and sports items	Toys

For the management of wastes, the "extended producer responsibility (EPR)" has been propagated as a latest paradigm, which is defined by the OECD as an approach of environmental policy in which the responsibility of producers for any product is extended to the post-consumer-stage of the life-cycle of products including its final-disposals (OECD, 2001).

There are a number of approaches to implement the EPR instruments both from administrative as well as legal perspectives varying from mandatory to fully-voluntary (OECD, 2001), as illustrated in Table 5.3.

TABLE 5.3 EPR Approaches with Examples (OECD, 2001)

EPR Approach Type	Examples
Taking-back programs for products	• Taking-back mandatorily • Taking-back voluntarily
Regulatory-approaches	• Minimizing standards of products • Prohibiting some hazardous products and materials • Disposing bans • Recycling mandatorily
Economic-instruments	• Schemes for deposits and refunds • Advanced recycling-fees • Disposal related fees • Taxes and subsidies on materials
Voluntary-industrial practices	• Performing codes of practices voluntarily • Private-public partnerships

5.1.2 MCDM APPROACH FOR E-WASTE MANAGEMENT

The use of MCDM techniques could be considered as beneficial methods in the E-waste management with conflicting diverse criterion (Kim et al., 2013; Rousis et al., 2008; Shumon et al., 2016; Queiruga et al., 2008). Since decision-making problems are mainly characterized by fuzziness, hence the FAHP technique has been developed to address the issues related to e-waste management in a study (Kahraman, 2008). MCDM methods include several techniques for supporting decisions that involves conflicting criterion, disproportionate variables, and a number of possible alternatives or solutions (Doumpos and Zopounidis, 2014; Pohekar and Ramachandran, 2004). The MCDM approaches are normally classified in two main streams, such as "multi-attribute decision making (MADM)" and "multi-objective decision making (MODM)." The selection or ranking problems are dealt in MADM, whereas optimization problems are performed using MODM. It may be noted that MADM methods aim in comparison or ranking of any set of alternatives on the basis of adopted criteria, while MODM techniques focus on identifying the set of optimal alternatives in accordance with the considered criteria. In a study in order to evaluate the barriers associated with the development of landfill communities, the "interpretive structural modeling-ISM" methodology (Chandramowli et al., 2011). Because of education and monthly incomes, most of the consumers do not receive payments for the recycling of electronic products (Yin et al., 2014). Ravi (2015) has used the ISM methodology to examine the interaction among the barriers of eco-efficiency in industries associated with electronic packaging. Welfens et al. (2016) have used the ISM approach to analyze the key enablers as well as barriers affecting the recurring and recycling of mobile phones. Pasalari et al. (2019) have used multi criteria decision making (MCDM) method and Fuzzy memberships in GIS environment to find out the best landfill sites for Shiraz county (south of Iran). The 15 most common sub-criteria considered in their study included groundwater, surface water, soil type, land use, distance to well, slopes, protected area, faults in environmental groups, residential areas, roads, villages, airport, infrastructures, wind direction in socio-economical group, and historical areas, respectively. It was concluded that the distances to residential areas as well as groundwater's were the most important criteria for landfill site selection, the six suitable areas for the landfill in Shiraz county was 1.003% of total area, and AHP and Fuzzy memberships having a great ability and potential for selection of landfill sites. A multi-criteria approach in order to select and rank the most suitable logistic service provider was carried out by the use of a Fuzzy-TOPSIS approach (Bottani and Rizzi, 2006). An integrated fuzzy technique for TOPSIS and multi-choice goal programming (MCGP) approach has been proposed in a study to solve the supplier selection problem in the supply chain (Liao and Kao, 2011). Zaeri et al. (2011) have proposed a methodology based on the TOPSIS approach for evaluating the suppliers in supply chain cycles. Rostamzadeh et al. (2013) have used a fuzzy multiple attribute decision making approach and TOPSIS method to assess and prioritize the cost-effectiveness criteria in supply chain management. Sohani and Sharma (2013) have used both AHP as well as TOPSIS for analyzing the performances of supply chain under an uncertain environment. Roghanian et al. (2014) have proposed a Fuzzy-TOPSIS approach for improving the supply chain management process in food industries. Nyaoga et al. (2016) have used the Grey-TOPSIS approach in tea processing firms for evaluating value chain performances. By the use of distance measures, the TOPSIS method has been used for ranking as well as selections of different alternatives. Zulqarnain and Dayan (2017) have proposed a Fuzzy-TOPSIS technique for decision makings. Therefore, on the basis of the positive outcomes of the MCDM method, the Fuzzy-TOPSIS method was used in the present analysis for the execution of proper selection of e-management strategies in India.

5.2 METHODOLOGY

In this study, five experts from environmental as well as electronic industrial sectors were considered to give their opinions. Then, the Fuzzy-TOPSIS method was used for the selection of the most suitable e-waste management strategies based on the opinion of experts and the result of analysis. The steps followed in this study was as illustrated in Figure 5.3.

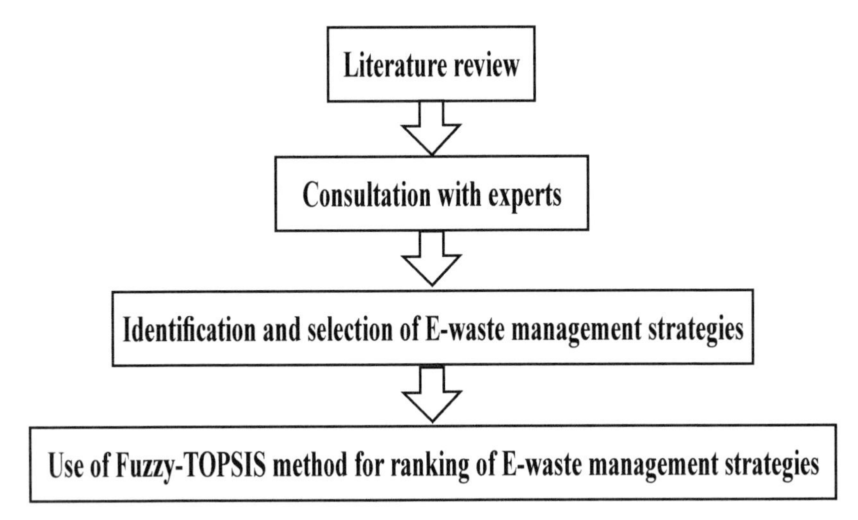

FIGURE 5.3 The steps followed by the Fuzzy-TOPSIS method.

The Fuzzy-TOPSIS method consisted of the following steps as illustrated below:

- **Step 1:** In matrix format a typical fuzzy multiple attribute group decision-making problem was constructed as:

$$\begin{array}{c} \quad\quad X_1 \ldots\ldots\ldots X_2, \ldots\ldots X_n \\ \hat{D} = \begin{array}{c} A_1 \\ A_2 \\ \vdots \\ A_M \end{array} \begin{bmatrix} \tilde{x}_{11} & \tilde{x}_{12} & \cdots & \tilde{x}_{1n} \\ \tilde{x}_{21} & \tilde{x}_{22} & \vdots & \tilde{x}_{2n} \\ \vdots & \vdots & \ddots & \vdots \\ \tilde{x}_{m1} & \tilde{x}_{m2} & \vdots & \tilde{x}_{mn} \end{bmatrix} \end{array}$$

$$\tilde{W} = (\tilde{W}_1, \tilde{W}_2., \tilde{W}_n)$$

where; $A_1, A_2, \ldots, A_m$ denoted the possible alternatives to be selected, X_1, $X_2 \ldots, X_n$ denoted the evaluation attributes (criteria) measuring the alternatives performances and correspond to the fuzzy performance rating of the ith alternative A_i versus the jth attribute X_j and $\tilde{W}_j$ denoted the attribute weight of X_j. Here, $\bar{x}_{ij}$;$\bar{\bar{x}}_{ij}, \forall_i, j$ = 1; 2;::: ; n were assessed in linguistic terms that were described by triangular fuzzy numbers, i.e.:

$$\bar{x}_{ij} = (a_{ij}, b_{ij}, c_{ij}),\ \tilde{W}_j = (\tilde{W}_{j1}, \tilde{W}_{j2}, \tilde{W}_{j3}).$$

- **Step 2:** In order to evaluate the important weights of the attributes, a group of k experts was established such as:

$$E = (E_1, E_2., E_k)$$

In addition, each group member was assigned with different voting power weights according to their professional titles as:

$$\bar{\lambda} = (\bar{\lambda}_1, \bar{\lambda}_2, \ldots \bar{\lambda}_k)$$

where; $\tilde{\lambda}_t$ represented the voting power weights of the t_{th} decision-maker that was expressed by triangular fuzzy numbers.

- **Step 3:** The fuzzy group opinion matrix by all experts can be expressed as:

$$\tilde{w}^* = \begin{matrix} & X_1, \ldots\ldots X_2, \ldots\ldots\ldots X_n \\ \begin{matrix} E_1 \\ E_2 \\ \vdots \\ E_M \end{matrix} & \begin{bmatrix} \tilde{w}^*_{11} & \tilde{w}^*_{12} & \cdots & \tilde{w}^*_{1n} \\ \tilde{w}_{21} & \tilde{w}^*_{11} & \vdots & \tilde{w}^*_{11} \\ \vdots & \vdots & \ddots & \vdots \\ \tilde{w}^*_{k1} & \tilde{w}^*_{k2} & \vdots & \tilde{w}_{Kn} \end{bmatrix} \end{matrix}$$

where; $\tilde{w}^*_{tj}$ indicated the fuzzy weight by the tth evaluator to the jth assessed attributes.

- **Step 4:** The following equation was used to integrate all the experts' opinions in order to aggregate the subjective judgment of k experts to obtain the fuzzy weight ($\tilde{W}_j$) of criteria (X_j).

$$\tilde{W} = \sum_{t=1}^{k} \tilde{\lambda}_t \otimes \tilde{w}^*_{tj},\ j = 1, 2, \ldots, n$$

- **Step 5:** Then, by applying the linear scale transformation method, the normalization of fuzzy decision matrix was performed as it preserves the property that the values of converted triangular fuzzy numbers remain in the range [0, 1]. Thus, the normalized fuzzy decision matrix ($\tilde{R}$) could be identified as:

$$\tilde{R} = |\tilde{r}|\ \mathrm{m} \times \mathrm{n},\ \mathrm{i} = 1, 2, \ldots \mathrm{m};\ \mathrm{j} = 1, 2, \ldots \mathrm{n}$$

$$\tilde{r}_{ij} = (a_{ij} / c_j^+, b_{ij} / c_j^+, c_{ij} / c_j^+)\ \mathrm{j} \in J^+$$

$$(a_j^- / c_{ij}, a_j^- / b_{ij}, a_j^- / c_{ij}),\ \mathrm{j} \in J^-$$

where; $\{c_j^+ = \max_i c_{ij}, a_j = \max_i a_{ij,} J^+\}$ was associated with benefit or maximization attributes or criteria and J^- was associated with cost or minimization attributes or criteria.

- **Step 6:** Then by multiplying the normalized fuzzy decision elements with the aggregative fuzzy weights of each criterion, the weighted normalized fuzzy decision matrix $\tilde{v}$ can be computed as:

$$\tilde{v} = |\tilde{v}_{ij}|\ \mathrm{m} \times \mathrm{n},\ \mathrm{i} = 1, 2, \ldots \mathrm{m},\ \mathrm{j} = 1, 2, \ldots \mathrm{n}.$$

where; $\max_i \tilde{v}_{ij} = \tilde{w}_j \otimes \tilde{r}_{ij}$, and $\tilde{v}_{ij}, \forall_i$ j were positive triangular fuzzy numbers.

- **Step 7:** Further, the fuzzy positive ideal solution (FPIS, A^+), as well as the fuzzy negative ideal solution (FNIS, A^-) were obtained as:

$$A^+ = \{(\max_i \tilde{v}_{ij} \mid \mathrm{j} \in J^+), (\min_i \tilde{v}_{ij} \mid \mathrm{j} \in J^-)\}$$

$$= \{\tilde{v}_1^+, \tilde{v}_2^+ \ldots, \tilde{v}_j^+, \ldots \tilde{v}_n^+,\}$$

$$A^- = \{(\min_i \tilde{v}_{ij} \mid \mathrm{j} \in J^+), (\max_i \tilde{v}_{ij} \mid \mathrm{j} \in J^-)\}$$

$$= \{\tilde{v}_1^-, \tilde{v}_2^-, \ldots, \tilde{v}_j^-, \ldots \tilde{v}_n^-\}$$

By considering the ranges of decision elements ($\tilde{v}_{ij}, \forall_i$ j) belonging to the closed interval [0, 1] satisfied that:

$$[\tilde{v}_{I+}^+ = \tilde{v}_{I-}^- = (1, 1, 1)], \text{ and } [\tilde{v}_{j+}^- = \tilde{v}_{j-}^+ = \tilde{v}_{j-}^+ = (0; 0; 0)].$$

where; J^+ was associated with benefit criteria, and J^- was associated with cost criteria.

- **Step 8:** Then, the distances of each alternative from (A^+) and (A^-) were obtained by the "Euclidean Distance Method" as:

$$S_i^+ = \sqrt{\sum_{j=1}^n d^2(\tilde{v}_{ij}, \tilde{v}_j^+)}, i = 1, 2, \ldots \mathrm{m}$$

$$S_i^- = \sqrt{\sum_{j=1}^n d^2(\tilde{v}_{ij}, \tilde{v}_j^-)}, i = 1, 2, \ldots \mathrm{m}$$

where; d ($\tilde{v}_A$, $\tilde{v}_B$) denoted the measured distance between two triangular fuzzy numbers, i.e., $\tilde{A}$ and $\tilde{B}$.

- **Step 9:** After calculating the S_i^+ and S_i^- of each alternative, a closeness coefficient (c_i^*) was calculated to determine the final ranking order of all alternatives as:

$$c_i^* = S_i^- / (S_i^+ + S_i^-), 0 < c_i^* < 1$$

It may be noted that as (c_i^*) approaches to 1, when the alternative (A_i) gets closer to (A^+) and gets farther from (A^-), respectively. Therefore, according to the corresponding closeness coefficients, the ranking order of all alternatives can be obtained.

5.3 RESULTS AND DISCUSSION

Based on the literature and opinion of experts,' the problems associated with the e-waste management system in India were considered as the seven criteria as illustrated in Table 5.4. The criteria to be maximized were taken as 1, while the criteria to be minimized as –1 for further analysis. Similarly, the possible solutions for the elimination of such problems as eight alternative measures were considered (Table 5.5).

TABLE 5.4 Selection of Criterion Concerned to E-Waste

Criteria	Description	Type of Criteria
X_1	Lower awareness level of hazards about the incorrect e-waste disposal among manufacturers and consumers.	Maximization
X_2	In-accurate estimation of the quantity of generated e-waste in India.	Maximization
X_3	In-accurate estimation of the quantity of recycled e-waste in India.	Maximization
X_4	Lack of knowledge about toxins in e-waste by the workers that result to health hazards.	Maximization
X_5	Processing of e-waste by the informal sectors resulting in severe environmental damages.	Minimization
X_6	In-effective recycling processes resulting in significant losses of material values as well as resources.	Minimization
X_7	Lack of specific legislation to deal with e-waste disposal, recycling, and management.	Maximization

The linguistic scale used for rating of the alternatives as defined by Chen and Hwang (1992) was as shown in Table 5.6.

The integrated matrix was obtained between the criterion and alternatives, with the corresponding weights (W_i) of all the seven criteria as shown in Table 5.7, which was followed by the formulation of normalized matrix (Table 5.8), and weighted normalized matrix (Table 5.9).

TABLE 5.5 Proposed Alternatives

Proposed Alternatives	Symbols of Alternatives
Restriction and domestic legal frameworks on importing of e-waste.	A_1
Addressing on the safe disposal of e-wastes both internal and external to domestic levels.	A_2
Attracting investments in e-waste management sectors.	A_3
Linking up the informal sectors activities with the formal sectors.	A_4
Adopting consultative process for an effective e-waste management plan.	A_5
Promoting of awareness program on recycling and disposal of e-waste.	A_6
Providing incentives and subsidized schemes to the general public or industries associated with recycling and disposal of e-waste.	A_7
Imposing hazardous e-waste disposal fees from manufacturers and consumers.	A_8

TABLE 5.6 Linguistic Scale with Corresponding Triangular Fuzzy Numbers

Linguistic Scale	Symbol	Triangular Fuzzy Numbers
Very poor	VP	[0, 0, 1]
Poor	P	[0, 1, 3]
Rather poor	RP	[1, 3, 5]
Fair	F	[3, 5, 7]
Rather good	RG	[5, 7, 9]
Good	G	[7, 9, 10]
Very good	VG	[9, 9, 10]

TABLE 5.7 Obtained Integrated Matrix

	W_1			W_2			W_3			W_4			W_5			W_6			W_7		
	0.56	0.93	0.98	0.77	0.93	1	0.77	0.93	1	0.54	0.65	0.74	0.33	0.63	0.73	0.44	0.55	0.77	0.77	0.87	0.97
	X_1			X_2			X_3			X_4			X_5			X_6			X_7		
A_1	14.80	20.20	24.60	2.400	6.400	11.80	2.800	6.400	12.00	6.800	12.00	17.80	6.800	12.00	17.80	13.20	19.60	25.80	5.000	9.000	13.60
A_2	3.800	8.200	13.80	10.80	16.60	22.20	16.20	21.80	26.00	15.00	20.80	25.40	15.00	20.80	25.40	15.00	20.80	25.40	14.40	19.80	24.00
A_3	5.800	10.80	16.60	3.400	7.600	13.00	3.200	7.400	13.00	2.400	6.400	11.80	2.400	6.400	11.80	2.400	6.400	11.80	2.400	6.400	11.80
A_4	3.000	6.800	12.40	19.00	24.40	28.00	17.00	22.60	26.80	19.40	24.60	28.00	17.80	23.00	26.40	19.40	24.60	28.00	19.40	24.60	28.00
A_5	13.80	19.80	24.60	6.800	12.20	18.20	3.600	8.000	14.20	2.800	7.000	12.60	2.800	7.000	12.60	2.800	7.000	12.60	2.800	7.000	12.60
A_6	2.000	5.600	11.00	2.400	5.800	11.00	12.60	18.60	24.00	8.200	13.80	19.60	8.200	13.80	19.60	8.200	13.80	19.60	8.200	13.80	19.60
A_7	16.60	22.20	26.40	1.800	4.800	9.600	3.600	8.200	13.80	2.600	5.800	10.80	2.600	5.800	10.80	2.600	5.800	10.80	2.600	5.800	10.80
A_8	6.000	11.20	17.00	6.800	12.20	18.20	7.600	13.00	18.80	4.400	9.000	14.60	4.400	9.000	14.60	4.400	9.000	14.60	4.400	9.000	14.60
Max	16.60	22.20	26.40	19.00	24.40	28.00	17.00	22.60	26.80	19.40	24.60	28.00	17.80	23.00	26.40	19.40	24.60	28.00	19.40	24.60	28.00
Min	2.000	5.600	11.00	1.800	4.800	9.600	2.800	6.400	12.00	2.400	5.800	10.80	2.400	5.800	10.80	2.400	5.800	10.80	2.400	5.800	10.80

TABLE 5.8 Obtained Normalized Matrix

	W_1			W_2			W_3			W_4			W_5			W_6			W_7		
	0.56	0.93	0.98	0.77	0.93	1	0.77	0.93	1	0.54	0.65	0.74	0.33	0.63	0.73	0.44	0.55	0.77	0.77	0.87	0.97
	X_1			X_2			X_3			X_4			X_5			X_6			X_7		
A_1	0.560	0.909	1.482	0.085	0.262	0.621	0.104	0.283	0.706	0.243	0.487	0.917	0.134	0.483	1.588	0.093	0.296	0.818	0.178	0.366	0.701
A_2	0.144	0.369	0.831	0.385	0.680	1.168	0.604	0.964	1.529	0.535	0.845	1.309	0.094	0.278	0.720	0.094	0.278	0.720	0.514	0.805	1.237
A_3	0.219	0.486	1.000	0.121	0.311	0.684	0.119	0.327	0.764	0.085	0.260	0.608	0.203	0.906	4.500	0.203	0.906	4.500	0.085	0.260	0.608
A_4	0.113	0.306	0.747	0.678	1.000	1.473	0.634	1.000	1.576	0.693	1.000	1.443	0.091	0.252	0.606	0.085	0.235	0.556	0.693	1.000	1.443
A_5	0.522	0.892	1.482	0.243	0.500	0.958	0.134	0.354	0.835	0.100	0.284	0.649	0.190	0.828	3.857	0.190	0.828	3.857	0.100	0.284	0.649
A_6	0.075	0.252	0.662	0.085	0.237	0.579	0.470	0.823	1.411	0.293	0.561	1.010	0.122	0.420	1.317	0.122	0.420	1.317	0.293	0.561	1.010
A_7	0.628	1.000	1.590	0.064	0.196	0.505	0.134	0.362	0.811	0.093	0.235	0.556	0.222	1.000	4.153	0.222	1.000	4.153	0.093	0.235	0.556
A_8	0.227	0.504	1.024	0.243	0.500	0.958	0.283	0.575	1.106	0.157	0.366	0.752	0.164	0.644	2.454	0.164	0.644	2.454	0.157	0.366	0.752

TABLE 5.9 Obtained Weighted Normalized Matrix

	X_1			X_2			X_3			X_4			X_5			X_6			X_7		
A_1	0.314	0.846	1.452	0.066	0.244	0.621	0.080	0.263	0.706	0.131	0.317	0.679	0.044	0.304	1.159	0.040	0.162	0.630	0.137	0.318	0.680
A_2	0.080	0.343	0.814	0.297	0.632	1.168	0.465	0.897	1.529	0.289	0.549	0.969	0.031	0.175	0.525	0.041	0.153	0.554	0.396	0.700	1.200
A_3	0.123	0.452	0.980	0.093	0.289	0.684	0.092	0.304	0.764	0.046	0.169	0.450	0.067	0.571	3.285	0.089	0.498	3.465	0.066	0.226	0.590
A_4	0.063	0.285	0.732	0.522	0.930	1.473	0.488	0.930	1.576	0.374	0.650	1.068	0.030	0.159	0.443	0.037	0.129	0.428	0.533	0.870	1.400
A_5	0.292	0.829	1.452	0.187	0.465	0.958	0.103	0.329	0.835	0.054	0.185	0.480	0.063	0.522	2.815	0.083	0.455	2.970	0.077	0.247	0.630
A_6	0.042	0.234	0.649	0.066	0.221	0.579	0.362	0.765	1.411	0.158	0.364	0.747	0.040	0.264	0.961	0.054	0.231	1.014	0.225	0.488	0.980
A_7	0.352	0.930	1.558	0.049	0.183	0.505	0.103	0.337	0.811	0.050	0.153	0.412	0.073	0.630	3.032	0.097	0.550	3.198	0.071	0.205	0.540
A_8	0.127	0.469	1.003	0.187	0.465	0.958	0.218	0.535	1.106	0.085	0.237	0.557	0.054	0.406	1.791	0.072	0.354	1.890	0.121	0.318	0.730

Then, the $M(v_{ij})$ values were obtained for the alternatives as shown in Table 5.10, which was followed by the calculation of the FPIS and FNIS, i.e., (A^+) and (A^-) values (Table 5.11).

TABLE 5.10 Calculated $M(v_{ij})^*$ Values

	X_1	X_2	X_3	X_4	X_5	X_6	X_7
A_1	0.870	0.310	0.349	0.375	0.502	0.278	0.378
A_2	0.413	0.699	0.964	0.602	0.244	0.249	0.765
A_3	0.518	0.355	0.387	0.221	1.307	1.351	0.294
A_4	0.360	0.975	0.998	0.697	0.210	0.198	0.934
A_5	0.858	0.536	0.422	0.239	1.133	1.169	0.318
A_6	0.308	0.288	0.846	0.423	0.422	0.433	0.564
A_7	0.947	0.246	0.417	0.205	1.245	1.282	0.272
A_8	0.533	0.536	0.619	0.293	0.750	0.772	0.389
Max	0.947	0.975	0.998	0.697	1.307	1.351	0.934
Min	0.308	0.246	0.349	0.205	0.210	0.1986	0.272

$^*M(v_{ij}) = [(-a_{ij}^2 + c_{ij}^2 - a_{ij}\, b_{ij} + c_{ij}\, b_{ij})/\{3(-a_{ij} + c_{ij})\}]$.

Tables 5.12 and 5.13 illustrated the measured distances d^+ and d^- values between two triangular fuzzy numbers, i.e., $\tilde{A}$ and $\tilde{B}$. Similarly, the corresponding S^+ and S^- values for all the alternatives were as shown in Table 5.14.

The final ranking of alternatives (Table 5.15) based on the closeness coefficient values indicated that "linking up the activities of the informal sector with the formal sectors" ranked first, which was followed by "addressing on the safe disposal of EW both internal and external to domestic levels," "imposing hazardous e-waste disposal fees from manufacturers and consumers," "adopting consultative process for an effective e-waste management plan," "promoting of awareness program on recycling and disposal of EW," "providing incentives and subsidized schemes to general public or industries associated with recycling and disposal of EW," "restriction and domestic legal frameworks on importing of EW," and "attracting investments in e-waste management sectors," respectively (Figure 5.4).

TABLE 5.11 Calculated A^+ and A^- Values

	X_1			X_2			X_3			X_4			X_5			X_6			X_7		
A^+	0.352	0.930	1.558	0.522	0.930	1.473	0.488	0.930	1.576	0.374	0.650	1.068	0.067	0.571	3.285	0.089	0.498	3.465	0.533	0.870	1.400
A^-	0.042	0.234	0.649	0.049	0.183	0.505	0.080	0.263	0.705	0.050	0.153	0.412	0.030	0.159	0.443	0.037	0.129	0.428	0.071	0.205	0.540

TABLE 5.12 Calculated (d^+) Values

	X_1	X_2	X_3	X_4	X_5	X_6	X_7
A_1	0.070	0.874	0.754	0.522	0.196	0.383	0.790
A_2	0.559	0.315	0.030	0.144	0.463	0.426	0.203
A_3	0.432	0.798	0.693	0.863	0	0	0.919
A_4	0.629	0	0	0	0.523	0.521	0
A_5	0.083	0.516	0.634	0.813	0.017	0.014	0.865
A_6	0.700	0.926	0.151	0.433	0.255	0.224	0.461
A_7	0	1.023	0.647	0.929	0.018	0.015	0.990
A_8	0.414	0.516	0.390	0.692	0.087	0.074	0.737

TABLE 5.13 Calculated (d^-) Values

	X_1	X_2	X_3	X_4	X_5	X_6	X_7
A_1	0.645	0.122	0	0.368	0.267	0.078	0.219
A_2	0.160	0.683	0.725	0.763	0.039	0.057	0.774
A_3	0.292	0.205	0.062	0.041	0.523	0.521	0.042
A_4	0.079	1.023	0.754	0.929	0	0	0.990
A_5	0.625	0.469	0.098	0.081	0.488	0.486	0.084
A_6	0	0.079	0.593	0.454	0.208	0.213	0.473
A_7	0.700	0	0.109	0	0.560	0.559	0
A_8	0.310	0.469	0.357	0.205	0.388	0.386	0.212

TABLE 5.14 Calculated S_i^+ and S_i^- Values

S_i^+	S_1^+	3.590	S_i^-	S_1^-	1.702
	S_2^+	2.141		S_2^-	3.204
	S_3^+	3.707		S_3^-	1.689
	S_4^+	1.673		S_4^-	3.776
	S_5^+	2.945		S_5^-	2.333
	S_6^+	3.151		S_6^-	2.022
	S_7^+	3.623		S_7^-	1.929
	S_8^+	2.912		S_8^-	2.331

TABLE 5.15 Ranking of Alternatives

Alternatives	Closeness-Coefficient Value	Ranking
A_1	0.322	7
A_2	0.599	2
A_3	0.313	8
A_4	0.693	1
A_5	0.442	4
A_6	0.391	5
A_7	0.348	6
A_8	0.445	3

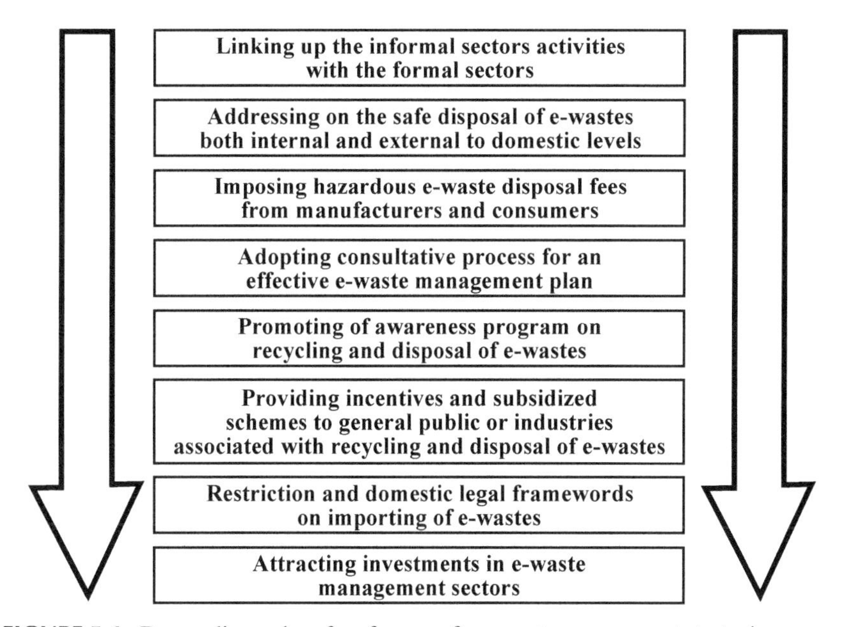

FIGURE 5.4 Descending order of preferences for e-waste management strategies.

5.4 CONCLUSION

In our daily lives, the new electronic products provide more comforts, security, easier, and quicker acquisition as well as exchange of information. On the other hand, these electronic products have also led to unrestrained consumption of resources and an alarming generation of hazardous wastes.

The e-waste management problems are faced by both the developed along with developing countries like India. One of the fastest-growing waste streams in the world consisting of EoL electrical and electronic products is primarily because of the rapid growth of technology, up-gradation in technical innovations and a higher rate of obsolescence in the electronic industries. Many of the electronic products and the generated EW contain toxic materials. Moreover, most of the trends in production and consumption processes are un-sustainable posing serious challenges to the environment as well as human health. Thus, the efficient and optimal use of natural resources, cleaner products development, wastes minimization, and environmentally sustainable recycling as well as disposal of wastes are some of the major issues that require attention by all concerns so as to ensure the economic growth of the country and thus by enhancing the quality of life.

KEYWORDS

- **barriers**
- **disposal**
- **electrical and electronic equipment**
- **electronic bonus card system**
- **e-waste**
- **fuzzy-TOPSIS**

REFERENCES

Anusree, P. S., & Balasubramanian, P., (2019). Awareness and disposal practices of e-waste with reference to household users in Kochi City. *International Journal of Recent Technology and Engineering, 8*(1S4), 293–299.

Arya, S., Gupta, A., & Bhardwaj, A., (2018). Electronic waste management approaches: A pilot study in Northern Indian States. *Int. J. Waste Resour., 8*(3), 346. doi: 10.4172/2252-5211.1000346.

Awasthi, A. K., Wang, M., Wang, Z., Awasthi, M. K., & Li, J., (2018). E-waste management in India: A mini-review. *Waste Management and Research, 36*(5), 408–414. doi: 10.1177/0734242X18767038.

Awasthi, A. K., Zeng, X., & Li, J., (2016a). Relationship between electronic waste recycling and human health risk in India: A critical review. *Environmental Science and Pollution Research, 23,* 11509–11532.

Awasthi, A. K., Zeng, X., & Li, J., (2016b). Environmental pollution of electronic waste recycling in India: A critical review. *Environmental Pollution, 211,* 259–270.

Babu, B. R., Parande, A. K., & Basha, C. A., (2007). Electrical and electronic waste: A global environmental problem. *Waste Management and Research, 25,* 307–318.

Bakhiyi, B., Gravel, S., Ceballos, D., Flynn, M. A., & Zayed, J., (2018). Has the question of e-waste opened a pandora's box? An overview of unpredictable issues and challenges. *Environ. Int., 110,* 173–192.

Bandyopadhyay, A., (2010). Electronics waste management: Indian practices and guidelines. *International Journal of Energy and Environment, 1*(5), 793–804.

Beigl, P. G., Wassermann, F., Schneider, & Salhofer, S., (2008). Forecasting MSW generation in major European cities. *European Commission's Fifth Frame Work Program (EVK4-CT-2002-00087)*, pp. 1–6.

Bhaskar, K., & Turaga, R. M. R., (2018). India's e-waste rules and their impact on e-waste management practices: A case study. *Journal of Industrial Ecology, 22*(4), 930–942.

Bhat, V., & Patil, Y., (2014). E-waste consciousness and disposal practices among residents of Pune city. *ICTMS-2013, Procedia-Social and Behavioral Sciences, 133,* 491–498.

Borthakur, A., & Govind, M., (2017). Emerging trends in consumers' e-waste disposal behavior and awareness: A worldwide overview with special focus on India. *Resources, Conservation and Recycling, 117,* 102–113.

Borthakur, A., & Singh, P., (2016). E-waste in the era of a digitally empowered India: A real challenge (A glance at the world). *Waste Manag., 58,* II.

Borthakur, A., & Singh, P., (2017). Researches on informal e-waste recycling sector: It is time for a 'lab to land' approach. *J. Hazard Mater., 323,* 730–732.

Borthakur, A., & Sinha, K., (2013). Electronic waste management in India: A stakeholder's perspective. *Electronic Green Journal, 1*(36), ISSN: 1076-7975.

Bottani, E., & Rizzi, A., (2006). A fuzzy TOPSIS methodology to support outsourcing of logistics services. *Supply Chain Management: An International Journal, 11*(4), 294–308.

Cairns, C. N., (2005). E-waste and the consumer: Improving options to reduce, reuse and recycle. *Proceedings of the IEEE International Symposium on Electronics and the Environment*, 237–242.

Chandramowli, S., Transue, M., & Felder, F. A., (2011). Analysis of barriers to development in landfill communities using interpretive structural modeling. *Habitat. Int., 35*(2), 246–253.

Chaturvedi, A., (2016). *E-Waste: An Introduction.* Retrieved from: www.adelphi.de (accessed on 18 February 2021).

Culver, J., (2005). *The Life Cycle of a CPU.* http://www.cpushack.net/life-cycle-of-cpu.html (accessed on 18 February 2021).

Davis, G., & Heart, S., (2007). Electronic waste: The local government perspective in Queensland, Australia. *Journal of Resources, Conservation and Recycling, 52*(8/9), 1031–1039.

Doumpos, M., & Zopounidis, C., (2014). *An Overview of Multiple Criteria Decision Aid: Multicriteria Analysis in Finance* (pp. 11–21). Berlin, Germany: Springer.

Dwivedy, M., & Mittal, R. K., (2012). An investigation into e-waste flows in India. *Journal of Cleaner Production, 37,* 229–242. doi: 10.1016/j.jclepro.2012.07.017.

EU, (2002a). Directive 2002/96/EC of the European Parliament and of the council of 27 January 2003 on waste electrical and electronic equipment (WEEE)-joint declaration of the European parliament, the council and the commission relating to article 9. *Official Journal L037:0024-39.* http://europa.eu.int/eur-lex/en/ (accessed on 18 February 2021).

EU, (2002b). Directive 2002/95/EC of the European Parliament and of the council of 27 January 2003 on the restriction of the use of certain hazardous substances in electrical and electronic equipment (RoHS). *Official Journal L037,* 19–23. http://europa.eu.int/eur-lex/en/ (accessed on 18 February 2021).

Ewasteguide, (2009). *E-Waste Definition.* Retrieved from Ewasteguide, website: http://ewasteguide.info/e_waste_definition (accessed on 18 February 2021).

Fowler, B. A., (2017). Magnitude of the global e-waste problem. In: *Electronic Waste Toxicology and Public Health Issues* (pp. 1–15). Amsterdam, The Netherland, Academic Press.

Frost and Sullivan Report, (2015). *Electronic Waste Management Services in India: A Perspective on Growth Opportunities,* Report No. 9835–15.

GTZ-MAIT, (2007). *E-Waste Assessment in India: Specific Focus on Delhi: A Quantitative Understanding of Generation, Disposal, and Recycling of Electronic Waste.* https://www.nswai.com/docs/e-Waste%20Assessment%20in%20India%20-%20Specific%20Focus%20on%20Delhi.pdf (accessed on 18 February 2021).

Gullett, B. K., Linak, W. P., Touati, A., Wasson, S. J., Gatica, S., & King, C. J., (2007). Characterization of air emissions and residual ash from open burning of electronic wastes during simulated rudimentary recycling operations. *Journal of Material Cycles and Waste Management, 9,* 69–79.

Hazardous Wastes (Management and Handling) Amendment Rules, (2003). *The Gazette of India Extraordinary: Part II, Section 3, Sub Section (ii).* Published by Authority No. 471, New Delhi. Ministry of Environment and Forests Notification, New Delhi.

He, W., Li, G., Ma, X., Wang, H., Huang, J., Xu, M., & Huang, C., (2006). WEEE recovery strategies and the WEEE treatment status in China. *Journal of Hazardous Materials, B136,* 501–512.

Jayapradha, A., (2015). Scenario of e-waste in India and application of new recycling approaches for e-waste management. *Journal of Chemical and Pharmaceutical Research, 7*(3), 232–238.

Kahraman, C., (2008). *Fuzzy Multi-Criteria Decision Making: Theory and Applications with Recent Developments.* Springer Science & Business Media, New York.

Kim, M., Jang, Y., & Lee, S., (2013). Application of Delphi-AHP methods to select the priorities of WEEE for recycling in a waste management decision-making tool. *J. Environ. Manag., 128,* 941–948.

Liao, C. N., & Kao, H. P., (2011). An integrated fuzzy TOPSIS and MCGP approach to supplier selection in supply chain management. *Expert Systems with Applications, 38*(9), 10803–10811.

Manga, V. E., Forton, O. T., & Read, A. D., (2008). Waste management in Cameroon: A new policy perspective. *Resources, Conservation and Recycling, 52,* 592–600.

Manomaivibool, P., (2009). Extended producer responsibility in a non-OECD context: The management of waste electrical and electronic equipment in India. *Resources, Conservation and Recycling, 53*(3), 136–144.

Mishra, S., Shamanna, B. R., & Kannan, S., (2017). Exploring the awareness regarding E-waste and its health hazards among the informal handlers in Musheerabad area of Hyderabad. *Indian J. Occup. Environ. Med., 21,* 143–148.

Monika & Kishore, J., (2010). E-waste management: As a challenge to public health in India. *Indian Journal of Community Medicine, 35*(3), 382–385.

Mundada, M. N., Kumar, S., & Shekdar, A. V., (2004). E-waste: A new challenge for waste management in India. *International Journal of Environmental Studies, 61*(3), 265–279. doi: 10.1080/0020723042000176060.

Murray, A., Skene, K., & Haynes, K., (2015). The circular economy: An interdisciplinary exploration of the concept and application in a global context. *J. Bus. Ethics, 140,* 369–380.

Nyaoga, R., Magutu, P., & Wang, M., (2016). Application of Grey-TOPSIS approach to evaluate value chain performance of tea processing chains. *Decision Science Letters, 5,* 431–446.

O'Connell, K. A., (2002). *Computing the Damage, Waste Age*. http://www.wasteage.com/ar/waste_computing_damage/ (accessed on 18 February 2021).

OECD, (2001). *Extended Producer Responsibility: A Guidance Manual for Governments*. Paris7 OECD.

Oliveira, C. R. D., Bernardes, A. M., & Gerbase, A. E., (2012). Collection and recycling of electronic scrap: A worldwide overview and comparison with the Brazilian situation. *Waste Management, 32,* 1592–1610.

Pasalari, H., Nodehi, R. N., Mahvi, A. H., Yaghmaeian, K., & Charrahi, Z., (2019). Landfill site selection using a hybrid system of AHP-Fuzzy in GIS environment: A case study in Shiraz city, Iran. *MethodsX, 6,* 1454–1466.

Peralta, G. L., & Fontanos, P. M., (2006). E-waste issues and measures in the Philippines. *Journal of Material Cycles and Waste Management, 8*(1), 34–39.

Pohekar, S. D., & Ramachandran, M., (2004). Application of multi-criteria decision making to sustainable energy planning: A review. *Renewable and Sustainable Energy Reviews, 8,* 365–381.

Pradhan, J. K., & Kumar, S., (2014). Informal e-waste recycling: Environmental risk assessment of heavy metal contamination in Mandoli industrial area, Delhi, India. *Environmental Science and Pollution Research, 21*(13), 7913–7928.

Puckett, J., & Smith, T., (2002). *Exporting Harm: The High-Tech Trashing of Asia The Basel Action Network.* Seattle7 Silicon Valley Toxics Coalition.

Queiruga, D., Walther, G., González-Benito, J., & Spengler, T., (2008). Evaluation of sites for the location of WEEE recycling plants in Spain. *Waste Manag., 28,* 181–190.

Rajya, S., (2011). *E-waste in India*. New Delhi: Rajya Sabha Secretariat. http://rajyasabha.nic.in/rsnew/publication_electronic/E-Waste_in_india.pdf (accessed on 18 February 2021).

Raut, E. R., Sudame, A. M., & Shanti, M., (2019). Scenario of e-waste in Nagpur city-a case study. *AIP Conference Proceedings, 2104,* 020037. Accessed from: https://doi.org/10.1063/1.5100405.

Ravi, V., (2015). Analysis of interactions among barriers of eco-efficiency in electronics packaging industry. *J. Cleaner Prod., 101,* 16–25.

Roghanian, E., Sheykhan, A., & Sayyad, A. E., (2014). An application of fuzzy TOPSIS to improve the process of supply chain management in the food industries: A case study of protein products manufacturing company. *Decision Science Letters, 3*(1), 17–26.

Rostamzadeh, R., Sabaghi, M., & Esmaili, A., (2013). Evaluation of cost-effectiveness criteria in supply chain management: Case study. *Advances in Decision Sciences,* 2013.

Rousis, K., Moustakas, K., Malamis, S., Papadopoulos, A., & Loizidou, M., (2008). Multi-criteria analysis for the determination of the best WEEE management scenario in Cyprus. *Waste Manag., 28,* 1941–1954.

Shagun, K. A., & Arora, A., (2013). Proposed solution of e-waste management. *International Journal of Future Computer and Communication, 2*(5), 490–493. doi: 10.7763/IJFCC.2013.V2.212.

Shevchenko, T., Laitala, K., & Danko, Y., (2019). Understanding consumer e-waste recycling behavior: Introducing a new economic incentive to increase the collection rates. *Sustainability, 11,* 2656. doi: 10.3390/su11092656.

Shumon, M., Ahmed, S., & Ahmed, S., (2016). Fuzzy analytical hierarchy process extent analysis for selection of end-of-life electronic products collection system in a reverse supply chain. *Proc. Inst. Mech. Eng. B J. Eng. Manuf., 230*(1), 157–168.

Sinha-Khetriwal, D., Kraeuchi, P., & Schwaninger, M., (2005). A comparison of electronic waste recycling in Switzerland and in India. *Environmental Impact Assessment Review, 25,* 492–504.

Sohani, N., & Sharma, P., (2013). Analyzing the performance of supply chain using TOPSIS. *International Journal of Science and Research, 2*(4), 221–224.

Tansel, B., (2017). From electronic consumer products to e-wastes: Global outlook, waste quantities, recycling challenges. *Environ. Int., 98,* 35–45.

Tran, P. H., Wang, F., Dewulf, J., Huynh, T. H., & Schaubroeck, T., (2016). Estimation of the unregistered inflow of electrical and electronic equipment to a domestic market: A case study on televisions in Vietnam. *Environ. Sci. Technol., 50*(5), 2424–2433.

Turaga, R. M. R., (2019). Public policy for e-waste management in India. *VIKALPA: The Journal for Decision Makers, 44*(3), 130–132.

UNEP, (1989). *Basel Convention on the Control of Transboundary Movements of Hazardous Wastes and Their Disposal,* United Nations Environment Program Secretariat of the Basel Convention. http://www.basel.int/text/documents.html (accessed on 18 February 2021).

Wath, S. B., Dutt, P., & Chakrabarti, T., (2011). E-waste scenario in India, its management and implications. *Environmental Monitoring and Assessment, 172*(1–4), 249–262.

Welfens, M. J., Nordmann, J., & Seibt, A., (2016). Drivers and barriers to return and recycling of mobile phones. Case studies of communication and collection campaigns. *J. Cleaner Prod., 132,* 108–121.

Yin, J., Gao, Y., & Xu, H., (2014). Survey and analysis of consumers' behavior of waste mobile phone recycling in China. *J. Cleaner Prod., 65,* 517–525.

Zaeri, M. S., Sadeghi, A., Naderi, A., & Kalanaki, A., (2011). Application of multi-criteria decision-making technique to evaluation suppliers in supply chain management. *African Journal of Mathematics and Computer Science Research, 4*(3), 100–106.

Zeng, X. L., Gong, R. Y., Chen, W. Q., & Li, J. H., (2016b). Uncovering the recycling potential of 'new' WEEE in China. *Environ. Sci. Technol., 50*(3), 1347–1358.

Zeng, X. L., Yang, C. R., Chiang, J. F., & Li, J. H., (2017). Innovating e-waste management: From macroscopic to microscopic scales. *Sci. Total Environ., 575,* 1–5.

Zeng, X., Duan, H., Wang, F., & Li, J., (2016a). Examining environmental management of e-waste: China's experience and lessons. *Renew Sustain. Energy Rev., 72,* 1076–1082.

Zulqarnain, M., & Dayan, F., (2017). Choose the best criteria for decision making via fuzzy TOPSIS method. *Mathematics and Computer Science, 2*(6), 113–119. doi: 10.11648/j.mcs.20170206.14.

CHAPTER 6

CHALLENGES IN THE MANAGEMENT OF BIOMEDICAL WASTES

ABSTRACT

Healthcare waste management is a crucial field for both practitioners and researchers. However, few studies have been conducted to evaluate it. This issue requires attention in different contexts. The unsafe practices known for waste management in developing countries like India raises questions about its effectiveness. Therefore, an attempt has been made in this study to identify the potential barriers hindering the waste management practices in Indian healthcare sectors by using the TOPSIS method. The significant barriers or challenges obtained were insufficient support from government agencies (ISFGA) followed by inadequate awareness and training programs (IAATP), financial constraints (FC), and Unauthorized Reuse of Health Care Waste (URHCW) that requires appropriate considerations and management.

6.1 INTRODUCTION

Bio-medical wastes or bio-wastes are the potentially hazardous waste materials that consist of liquids, solids, sharps, and laboratory-wastes. Bio-medical wastes differ from other hazardous wastes like industrial, electronic or agricultural wastes, as it belongs to biological sources or the residues from the prevention, diagnosis, or treatment of diseases. Hospitals and nursing homes are the major sources of bio-medical wastes. Normally the peoples from every walk of life in the society visits hospitals frequently and most of them produce wastes which are rising in its' amount as well as types owing to the advancement in scientific knowledge and thus also creating its' impacts. However, the public health and environment has been posted to a serious threat due to the hospital wastes in addition to the enriching risks for patients and personnel handling these wastes. Bio-medical waste management has become a significant issue for minimizing the potential health risks and damages to the environment. If these bio-medical wastes are not handled properly (Figure 6.1), then it may cause serious impacts on human well-being. So there is a prime requirement for the safer and proper disposal as well as treatment of these wastes through appropriate management practices.

FIGURE 6.1 Disposed medical wastes in hospital premises.

6.1.1 LITERATURE

A number of legislatures have been implemented in India, to ensure proper bio-medical wastes disposal that includes the "biomedical waste (management and handling) rules, (1998)," "hazardous wastes (management and handling) rules, (1989)," municipal solid wastes (MSWs) (management and handling) rules (2000), and so on. These rules are pertinent to every hospital as well as nursing-homes generating bio-medical wastes. Sufficient knowledge of health hazards of bio-medical wastes, and proper preventive measures in wastes handling can lead to safer hazardous-wastes disposal, and thus protects from the transmitted diseases. In spite of having a higher global awareness about hazardous-wastes among health care professionals and appropriate techniques for wastes management, the level of satisfaction of awareness in India is very low (Kishore et al., 2000; Pandit et al., 2005; Rao, 2008). Out of total existing beds in hospitals in India, the rural-hospitals are contributing to about 20% of total beds, while 80% are in urban-hospitals. Based on past figures on a number of beds in addition to the average quantity of waste generation rate at 1 kg per bed per day, it has been estimated that about 0.33 million-tons of hospital wastes being generated per year (Patil and Shekdar, 2001). An alarming situation for local governments in India has been created by the exponential growth in healthcare units like hospitals and dispensaries. Adverse effects on human health have been identified through the pollutants from these wastes. From medical waste incinerators, two detected pollutants having significant amount in air and ash emissions were revealed as mercury and dioxin. The burning of wastes openly by dispensaries, clinics, and hospitals lead in the release of dioxin to atmospheres (Kaiser et al., 2001). Very few studies have been dealt with the healthcare waste management practices and most of the studies have discussed about the harmful impacts of improper management of wastes and the Indian hospitals' status of current practices (Athavale and Dhumale, 2010; Gupta and Boojh, 2006; Gupta et al., 2009; Rao et al., 2004), and also very few studies have analyzed the issues and various barriers to waste management practices have been identified in Indian healthcare sectors (Patil and Shekdar, 2001; Verma, 2010). Past studies have revealed about waste management practices by overlooking the potential barriers inhibiting in the implementation of the waste management strategies in healthcare sectors. It has been observed of addressing on only the costs related directly to waste disposals including gathering, transport, handling, final-disposal of the wastes and efficient resources utilization by most of the health care administrators. However, these generated wastes have an indirect impact on human health as well as environment even after disposal also. Moreover, there is lacking of segregation practices in previous studies, and the mixing of hospital wastes with general waste makes the whole waste-stream to be harmful (Gupta and Boojh, 2006). If the solid medical wastes are improperly managed, then it can cause serious health risks and environmental problems. So, proper treatment and disposing of the medical wastes is very important (Mathur et al., 2011). Because of the infectious and hazardous nature of healthcare wastes causing undesirable effects on humans as well as the environment, the management of health care waste becomes a serious concern. The adoption of suitable strategies by healthcare units to manage these wastes have been forced through government regulations and growing public-awareness regarding healthcare waste issues. Even though various techniques are available for healthcare waste management as well as reduction, however the waste management practices in healthcare sectors are not free from challenges in developing countries like India. Muduli and Barve (2012a) have made an attempt for the identification of major challenges in healthcare waste management faced by the healthcare units in India. Owing to an increase in pollution-levels and varying lifestyles associated with rapid civilization, the healthcare services have become a basic need for peoples irrespective of their gender, age, and culture. Muduli and Barve (2012b) have identified the potential barriers hindering the greening effort of the healthcare waste sectors in India, and used the "interpretive structural modeling" for modeling as well as analyzing the identified barriers. The important steps in appropriate bio-medical waste management process include segregation, storages, transportations, treatments, and disposal (Asadullah et al., 2013; Singh et al., 2014), that need for special attention (Chartier et al., 2014; FMHACA, 2005; FmoH, 2008; Singh et al., 2014). Kumari et al. (2013) have suggested the features and step-by-step approaches to establish a bio-medical waste management system in hospitals at tertiary levels and to provide hands-on-training to all "State Medical Colleges" and other hospitals to establish an effective bio-medical waste management system. Sarker et al. (2014) have made a study among healthcare providers to assess the knowledge and practices regarding medical waste management for identifying possible associated barriers in Dhaka Division (Bangladesh). They collected data from 625 healthcare providers, including 220 nurses, 245 medical doctors, 116 cleaning-staffs, and 44 technologists, directly involved in medical waste management by the use of self-administered as well as semi-structured questionnaires. It was concluded that in order to improve knowledge and practices regarding medical waste management, the expansion and strengthening of ongoing educational training/programs to be essential, and the government should take necessary steps and make available of financial supports for eliminating the possible associated barriers to properly manage the medical wastes. Windfeld and Brooks (2015) have carried out a review focusing on healthcare wastes management practices, and sorted them into grouping and separations, transportation as well as disposal, and presented the regulatory-practices of the UK, Canada, the EU, the US, and other developing nations. Holla et al. (2015) have tried to determine the knowledge about the proper biomedical waste disposal of healthcare professionals, and their practices in following the preventive measures in handling the biomedical wastes, and found the awareness to be better among staff nurses, lab-technicians, and doctors, as compared to class-four employees.

The medical wastes are referred to the waste's generation during the process of diagnosis, treatments, operations, or immunization (Nagaraju et al., 2013; Ismail et al., 2013). It has been revealed of 75–95% of bio-medical wastes to be non-hazardous, while the remaining 10–25% as hazardous to animals and humans in addition to the environment (Askarian et al., 2004; Bhatt et al., 2013; Ozder et al., 2013). It may be noted that the combination of these make these wastes more harmful (Singh et al., 2007). Different studies have shown of 80% of all medical-wastes getting mixed with general-wastes (Ruoyan et al., 2010). As per the estimation of WHO during 2000, the injections with contaminated syringes caused human immunodeficiency virus (HIV) infections of 260 000 cases, hepatitis-B-virus (HBV) infections of 21 million cases, 2 million hepatitis-C-virus infections (Shinee et al., 2008). After needle injury, the cases of "staphylococcal bacteriemia along with endocarditis" were reported among cleaning-staff (Sachan et al., 2012). As the health-care providers perform their jobs in hospitals, they are always at greater risks of occupational-dangers (Battle, 1994). The waste generation-rate from government hospitals were reported to be 0.11 and 0.03 kg per bed per day for infectious and sharps wastes, respectively (Directorate General of Health Services (Bangladesh), 2005). The medical wastes are found to be mixed with the municipal-wastes in the collecting-bins at the roadsides with some percentage buried without any safety measures or burned openly (Hassan et al., 2008). Even though in order to manage the bio-medical wastes, appropriate steps are undertaken by "The Ministry of Health and Family Welfare, Bangladesh," but still satisfactory results are not achieved (Hassan et al., 2008; Directorate General of Health Services (Bangladesh), 2012), and adverse health effects are reported because of medical-wastes (Akter and Trankler, 2003). However, the WHO has recommended to raise awareness regarding medical-wastes associated risks and promoting safe practices to improve the circumstances (World Health Organization (WHO)). As compared to medical-doctors and nurses, the knowledge levels were revealed to be inadequate among technologists as well as cleaning staff (Sachan et al., 2012; Mostafa et al., 2009; Saini et al., 2005; Mathur et al., 2011; Pandit et al., 2005; Amanullah et al., 2008), the personal-protective-devices are also not utilized at most of the times by the cleaning staff (Biswas et al., 2011). It was reported in 1994, that inappropriate waste management was attributed because of the local healthcare professionals' negligence (Halbwachs, 1994). The role of hospital-nurses are not only to protect themselves, but also at the same time to reduce the associated risk exposures to other healthcare professionals, patients as well as attendants (Rahman et al., 2001).

Unregulated bio-medical waste management has threatened not only the human health and safety, but also to the environment in view of the present and future generations. The HBV outbreak incidence with 240 infections at Gujarat (India) in 2009, and the generation of mass-vaccinations (1.6 million) in Afghanistan due to bio-medical wastes (Chartier et al., 2014). Previous studies have estimated of about half of the population in the world to be at risk from hazards due to improper treatment and management of bio-medical wastes (Harhay et al., 2009). In a survey which was performed by "International Clinical Epidemiology Network" in 25 districts considering 20 states in India, the two big cities such as Chennai and Mumbai were found to have comparatively better bio-medical wastes management systems. Further, it was observed that about 82%, 60% and 54% of primary, secondary, and tertiary healthcare facilities were in the red-category requiring major improvements (IPEN, 2014). Studies conducted by the "WHO" in 22 developing countries have revealed of the proportions of healthcare facilities not using proper methods of waste disposals to be in the range of 18–64% (WHO, 2011). An annual amount of about 0.33 million-tons of bio-medical wastes ranging from 0.5 to 2.0 kg per-bed per-day has been reported to be generated in India (Mathur et al., 2011). The poor practices to manage bio-medical wastes are attributed because of lack of awareness as well as training (David and Shanbag, 2016). The bio-medical wastes rule were implemented at forest in India in 1998, that was amended as draft in 2003 and 2011 under the "Environment Protection Act (EPA)-1986" (The Gazette of India Biomedical Wastes (Management and Handling) Rules, 1998). India has been a participant of "Stockholm

Convention (2004)," to eliminate as well as restrict the production of "persistent organic pollutants (POPs)" (Secretariat of the Stockholm Convention, 2006). In order to fill the gaps in the old-rules for regulating the disposals of various categories of bio-medical wastes, the "Ministry of Environment Forests and Climate Change, Government of India," have notified the bio-medical wastes management rules on 28th March 2016, under the provisions of EPA, 1986 (Bio-Medical Waste Management Rules, 2016). All the hazardous healthcare wastes include sharp-wastes, infectious-wastes, pathological-wastes, pharmaceutical-wastes, cytotoxic-wastes, chemical-wastes, infectious liquid wastes, radioactive-wastes, and general healthcare wastes (WHO, 2004a). The "International Solid Waste Association" has been acting as an international, independent, and non-profit making association in order to develop and promote the professional as well as sustainable waste management throughout the world (Secretariat of the Stockholm Convention, 2006; WHO, 2004a, 2007). Similarly, in order to promote environmentally sound bio-medical wastes management, the WHO suggests of adopting poly-vinyl chloride (PVC) free medical-devices, recycling, risk-assessments, and sustainable technology (WHO, 2004a, 2007). As per the WHO guidelines, for proper management of bio-medical wastes, the laboratory waste including specimens of micro-organisms as well as infectious wastes of patients in isolation, need to be pre-treated on sites by the use of safer plastic containers or bags, and then sent for incineration to common biomedical waste management disposal facility (PA-DEP; National Research Council, 1989; WHO, 2004b). The radioactive wastes treatments as well as disposals are under "national nuclear regulatory agency" and its rules. For instance, the isotopes with longer half-life, it is recommended to have long-time storage at authorized waste disposal sites. For, infectious radioactive wastes, it is recommended to be decontaminated after containment by decay-time, which is usually ten-times the half-life, in an identified, isolated, and designated room sites before final disposals. However, the lower-level radioactive wastes can be discharged in sewers (IARC, 1985). Moreover, it has been recommended for cytotoxic-wastes including all contaminated items with cytotoxic-drugs to be put in a "non-chlorinated yellow-container," and sealed as well as labeled as cytotoxic. The expired cytotoxic-drugs need to be returned to the suppliers or manufacturers for subsequent incineration at temperature more than 1200°C (UN, 2010). All the bio-medical wastes are required to be labelled as waste-types, site of generation, generation-date before transporting from generation sites. As per the WHO guidelines, the central storage-site need to be cleaned once weekly, with all transport staffs wearing adequate "personal protective equipment." It should have adequate facilities for good drainage, an exhaust and water supply, and also the facility to keep general wastes detached from bio-medical wastes (Chartier et al., 2014; Facility Guidelines Institute, 2010; UN, 2009).

The significant differences between bio-medical wastes rules (1998) and bio-medical wastes management rules (2016) are as illustrated in Table 6.1.

Das and Biswas (2016) have assessed the knowledge as well as practices of hospital wastes management among healthcare providers of a tertiary-care hospital which was conducted in the sections of surgery, general medicines, gynecology, obstetrics, and radiotherapy among 198 hospital staffs during 3 months period by using a pre-designed and pre-tested interview schedule to bring out the bio-medical waste management knowledge. It was observed that 6.6% of staffs were known about five-color coding for segregation of wastes with different color bags like black, red, yellow, and blue, and white puncture-proofs containers. Moreover, 31.3% of staffs were known about correct sharps disposal, and everyone had awareness about the utilization of personal protective-measures during bio-medical wastes handling. Adequate-knowledge is a fundamental requirement for proper bio-medical waste management practices. However, a better result was obtained with 96% of the study participants having good knowledge scores in Pakistan (Ajmal and Ajmal, 2017). Different studies have indicated a lack of awareness, training, staff-resistances, managerial poor-commitments, negligence, inadequate resources, and un-favorable attitudes of the healthcare staffs as the major identified challenges (Demissie, 2014; Debalkie and Kumie, 2017; Doylo et al., 2018; Deress et al., 2018; Haylamicheal et al., 2011). Delmonico et al. (2018) have investigated the barriers of waste management healthcare sectors by analyzing the waste management practices in two Brazilian hospitals with the use of "Analytic-Hierarchy-Process" method. By organizing the barriers into three categories such as human-factors, management, and infrastructures, it was suggested that employee awareness and cost were the most important barriers. Hassan et al. (2018) have assessed the management issues related to the use of needles and suggested suitable advices for a safer and improved system management of needles in Khartoum (Sudan). It was revealed that the management of both home-generated healthcare wastes and healthcare in Sudan was deficient as all

TABLE 6.1 The Significant Differences between Bio-Medical Wastes Rules (1998) and Bio-Medical Wastes Management Rules (2016)

Parameters	Bio-Medical Waste Rules (1998)	Bio-Medical Waste Management Rules (2016)
Occupiers' duties	• Duties are not better defined • Absence of wastes pretreatment on sites • Recommendation of "Chlorinated-plastic bags, gloves, blood-bags" • "Effluent treatment plant (ETP)" is optional • Not compulsory maintenance of the record-details • Not necessarily to be posted the annual reports on the website • Non-compulsory BMWM committee • Not compulsory maintenance of the records	• Duties are better defined • Infectious lab-waste blood-bag's pretreatment by disinfection as well as sterilization on sites as per WHO guidelines. • Within 2-years of notification, the occupiers ensure of nonchlorinated-plastic bags, gloves, blood-bags. The occupiers also make sure of segregation of liquid-wastes by pretreatment at sources • Requirement of mandatory ETP • Occupiers make sure of daily maintenance of BMWM register and on monthly for a website • On the website, the annual-report have to be made available within 2 years • BMWM committee is established by the occupiers (more than or equal to 30 bedded) • Compulsory records of equipment, health-checkups, trainings, vaccinations
CBMWTFs' duties	• Duties are not better defined • Un-documented "Bar-coding, GPS, and vaccinations for HCWs" • Un-documented records	• Duties are better defined • Establishment of "Bar-coding, GPS, and vaccinations for HCWs" by the occupiers • Accidents' reporting and records-maintenance of equipment, health-checkups, trainings
Reporting of accidents	• No specific-reporting of accident's	• Reporting of major-accidents to authorities and to be retained in annual reports

TABLE 6.1 *(Continued)*

Parameters	Bio-Medical Waste Rules (1998)	Bio-Medical Waste Management Rules (2016)
Microbiology and biotechnology wastes	• Optional pretreatment	• Mandatory infectious-wastes' pretreatment as per WHO guidelines
Liquid-infected wastes	• It has been mentioned for chemical-treatment for liquid-wastes and discharge into drains, in order to conform to effluent-standards	• Mandatory effluent-treatment plant and effluents are required to conform to standards
Infected-plastics, sharps, and glasses	• Infected-plastics, metal-sharps, glasses are carried in blue-containers with disinfectants, and local-autoclaving or incineration or microwaving is recommended	• Infected-plastics, metal-sharps, glasses are carried in red-bags and white-containers, respectively. Then, these are sent to authorized-recyclers. • The glass-articles are discarded in cardboard-boxes with blue-marks
Recycling	• Absence of the mention of authorized-recyclers for recycling of plastics and glasses	• Focus on authorized-recyclers for recycling of plastics and glasses

the wastes were mixed together and improperly disposed, especially the used needles. The reason behind the negligence included many factors such as lacking in waste segregations at the source, lacking in policies, planning failures, in-adequate trainings, lacking in awareness of the hazardous wastes, weaker infrastructures, and lacking inappropriate treatment technologies. The transmission of more than 30 dangerous blood-borne pathogens occurs due to improper healthcare waste management. Lacking in awareness of the healthcare staff, suitable waste management utilities, and the regulatory bodies' enforcement were identified as a major general factors shared in most of the studies. Thus, there is a requirement of close supervision by the regulatory bodies or other stakeholders for the waste disposal processes (Yazie et al., 2019). Niyongabo et al. (2019) have studied the current practices of solid medical wastes management from collection to final disposal phase in twelve healthcare facilities of Burundi based on the official government reports. It was observed that for on-site or off-site solid medical wastes transportation, 75% and 92% of healthcare facilities used un-covered wheelbarrows as well as trucks, respectively, which indicated the un-safe protection of most transportation equipment and waste workers. Further, 15,736.4 tons of solid medical wastes, i.e., 92.8%, from all twelve healthcare facilities were improperly disposed through un-controlled land disposal as well as incineration.

6.1.2 INTERNATIONAL SCENARIO IN BIO-MEDICAL WASTES MANAGEMENT

About 18–64% of healthcare facilities have un-satisfactory bio-medical waste management facilities at the global level, that include the predictors as insufficient-resources, lack of awareness, and underprivileged disposal-mechanisms. It has been reported that about 56% of facilities in South-East Asian-region countries are lacking of appropriate disposal and treatment of waste (WHO, 2011). Different authors have reported of the similar situations in other developing countries like Nigeria, Pakistan, Iran, and Senegal, with poor infrastructures, state-of-collection, transportations, disposals, training, capacity-building, personal-protective-equipment, and resource-constraints for an effective management of bio-medical wastes (Askarian et al., 2004; Abah and Ohimain, 2011; Ali et al., 2015). The major gap which was found in many studies in India includes in adequate knowledge and practices in relation to resources-availability and processes involved, which need for an organized-training in addition to structured-supervision. It has been found through various studies in India that even though the people are with higher-education background on tertiary care-hospitals like residents, consultants, and scientists, but in actual practices, there was lacking of the reflection of good knowledge of bio-medical rules (Mathur et al., 2011; David and Shanbag, 2016; Bhagawati et al., 2015; Saini et al., 2005; Sarotra et al., 2016). Moreover, it was revealed from geographically diverse states of India that regarding segregation of bio-medical wastes, there was a higher awareness among hospital-staffs in urban-areas as compared to rural-areas (IPEN, 2014; UNIDO, 2010).

Moreover, the "technique for order preference by similarity to ideal-solution" (TOPSIS) was developed by Hwang and Yoon (1981) for evaluating the performances of alternatives' with the ideal solution through similarities. In accordance with this technique, the alternative closest to the positive-ideal solution and farthest from the negative-ideal solution is most preferable. The benefit criteria are maximized by the positive-ideal solution and the cost criteria are minimized, while the cost criteria are maximized by the negative-ideal solution and the benefit criteria are minimized. Moreover, all best values attainable of criteria belong to the positive-ideal solution, and all the worst values attainable of criteria belong to the negative-ideal solution. TOPSIS method has been extensively used in a variety of research studies (Wojciech, 2013). An approach based on TOPSIS was used for algorithm ranking that has been named as "A-TOPSIS" to solve the ranking problems and comparison of algorithms (Krohling and Pacheco, 2015). Sari et al. (2018) have used the TOPSIS method to determine the best suppliers' for supplying main raw-materials of fresh as well as frozen chicken-meats on the slaughterhouse chicken industries. In this context, the present study was aimed at assessing the practices and challenges of health care professionals in handling bio-medical wastes. Thus, an attempt was made here for the identification of the potential barriers hindering the waste management practices in the Indian healthcare sectors.

6.2 METHODOLOGY

To find the significant barriers in bio-medical waste management practices in the Indian context, an extensive review of literature was done, and suggestions were obtained from five numbers of experts' in the field of medical sectors as well as twelve numbers of workers dealing with cleaning and wastes disposal activities. Further, in order to rank the most significant barriers as the challenges in bio-medical wastes management, the "TOPSIS method" was used (Figure 6.2).

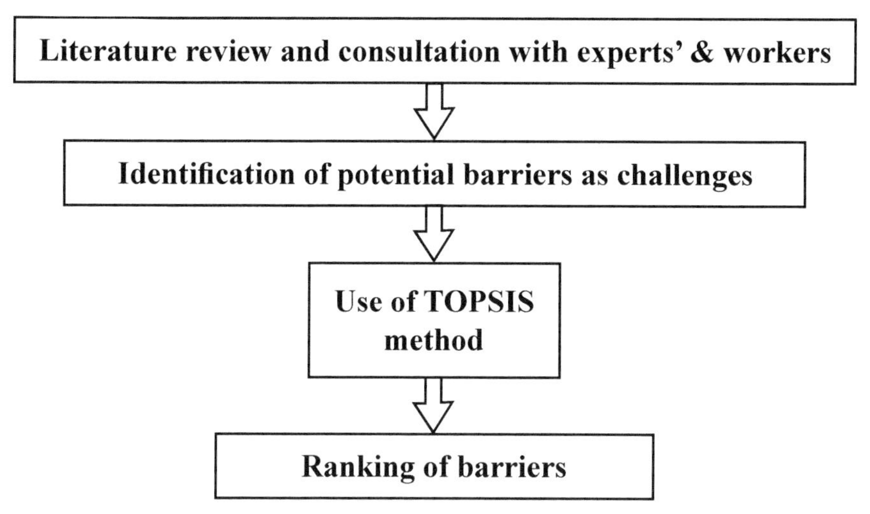

FIGURE 6.2 Flow chart of the present work.

The TOPSIS method utilizes the principle that the chosen alternative must be at closest and utmost distance from a geometric viewpoint by the use of the "Euclidean distance" for determining the "closeness coefficient" of an alternative with the optimal solution.

Generally, the TOPSIS method is based on the following steps:

- **Step 1:** Making a normalized decision matrix as in Eqn. (1):

$$r_{ij} = \frac{x_{ij}}{\sqrt{\sum_{i-1}^{m} x_{ij}^2}}, \text{i} = 1, ..., \text{m}; \text{j} = 1, ..., \text{n}, \quad (1)$$

- **Step 2:** Creating a weighted normalized decision matrix as in Eqn. (2);

$$y_{ij} = w_{ij}, \text{i} = 1, ..., \text{m}; \text{j} = 1, ..., \text{n} \quad (2)$$

- **Step 3:** Determining the matrix of positive ideal solutions (PIS) as well as the ideal negative solution (NIS) matrix as in Eqns. (3) and (4);

$$A^{+} = \left(y_1^{+}, y_2^{+}, \ldots\ldots, y_n^{+}\right) \tag{3}$$

$$A^{-} = \left(y_1^{-}, y_2^{-}, \ldots\ldots, y_n^{-}\right) \tag{4}$$

- **Step 4:** Determining the distance between the score for each alternative and the matrix of PIS and the NIS matrix as in Eqns. (5) and (6);

$$D_i^{+} = \sqrt{\sum_{j=1}^{n} \left(y_i^{+} - y_{ij}\right)^2} \tag{5}$$

$$D_i^{-} = \sqrt{\sum_{j=1}^{n} \left(y_{ij} - y_i^{-}\right)^2} \tag{6}$$

- **Step 5:** Determining the closeness coefficient (preference value) for each alternative as in Eqn. (7);

$$C_i = \frac{D_i^{-}}{D_i^{-} + D_i^{+}} \tag{7}$$

6.3 RESULTS AND DISCUSSION

For establishing a decision matrix for the ranking, the structure of the matrix can be expressed as in Table 6.2. The selected criteria were medical waste, biomedical waste, clinical waste, bio-hazardous waste, regulated medical waste, and infectious medical and healthcare waste, respectively. Similarly, the barriers selected through the literature, opinions of experts' and workers were as the followings:

1. Lack of segregation practices (LOSP): B_1;
2. Improper waste management operational strategy (IWMO): B_2;
3. Insufficient support from government agencies (ISFGA): B_3;
4. Lack of green procurement policy (LOGPP): B_4;
5. Unauthorized reuse of health care waste (URHCW): B_5;
6. Lack of top management commitment (LOTMC): B_6;
7. Lack of adequate facilities (LOAF): B_7;
8. Financial constraints (FC): B_8;
9. Inadequate awareness and training programs (IAATP): B_9;
10. Reluctance to change and adoption (RTCAA): B_{10}.

The normalized decision-matrix (Table 6.4) was obtained based on the decision-matrix (Table 6.3).

TABLE 6.2 Selecting Criteria for Health-Sector Waste Evaluation and Weight of Criteria

Code	Criterion	Weight (%)
C1	Medical waste	0.1%
C2	Biomedical waste	0.2%
C3	Clinical waste	0.1%
C4	Bio-hazardous waste	0.3%
C5	Regulated medical waste (RMW)	0.2%
C6	Infectious medical waste and healthcare waste	0.3%

TABLE 6.3 Calculating the Decision-Matrix

	B_1	B_2	B_3	B_4	B_5	B_6	B_7	B_8	B_9	B_{10}
C1	5	2	2	5	2	4	2	1	1	5
C2	4	1	2	4	4	1	3	4	2	1
C3	1	4	1	5	3	3	5	2	5	5
C4	4	5	5	2	1	2	4	1	5	2
C5	4	2	1	1	5	4	6	5	4	1
C6	4	1	1	4	4	5	2	4	4	5

TABLE 6.4 Calculating the Normalized Decision-Matrix

	C1	C2	C3	C4	C5	C6
B_1	0.227273	0.181818	0.045455	0.181818	0.181818	0.181818
B_2	0.133333	0.066667	0.266667	0.333333	0.133333	0.066667
B_3	0.166667	0.166667	0.166667	0.416667	0.083333	0.083333
B_4	0.238095	0.190476	0.238095	0.095238	0.047619	0.190476
B_5	0.105263	0.210526	0.157895	0.052632	0.263158	0.210526
B_6	0.210526	0.052632	0.157895	0.105263	0.210526	0.263158
B_7	0.090909	0.136364	0.227273	0.181818	0.272727	0.090909
B_8	0.058824	0.235294	0.117647	0.058824	0.294118	0.235294
B_9	0.047619	0.095238	0.238095	0.238095	0.190476	0.190476
B_{10}	0.263158	0.052632	0.263158	0.105263	0.052632	0.263158

The weighted normalized decision matrix was calculated (Table 6.5) by multiplying the normalized decision matrix by its associated weights, which was followed by the calculation of the PIS and NIS (Table 6.6).

TABLE 6.5 The Weighted Normalized Values

	C1	C2	C3	C4	C5	C6
B_1	0.022727	0.036364	0.004546	0.054545	0.036364	0.054545
B_2	0.013333	0.013333	0.026667	0.1	0.026667	0.02
B_3	0.016667	0.033333	0.016667	0.125	0.016667	0.025
B_4	0.02381	0.038095	0.02381	0.028571	0.009524	0.057143
B_5	0.010526	0.042105	0.01579	0.01579	0.052632	0.063158
B_6	0.021053	0.010526	0.01579	0.031579	0.042105	0.078947
B_7	0.009091	0.027273	0.022727	0.054545	0.054545	0.027273
B_8	0.005882	0.047059	0.011765	0.017647	0.058824	0.070588
B_9	0.004762	0.019048	0.02381	0.071429	0.038095	0.057143
B_{10}	0.026316	0.010526	0.026316	0.031579	0.010526	0.078947

TABLE 6.6 Determining the PIS and NIS

	C1	C2	C3	C4	C5	C6
A^+	0.026316	0.047059	0.026667	0.071429	0.058824	0.078947
A^-	0.004762	0.10526	0.00454	0.01579	0.009524	0.02

Then, the separation measures were calculated by using the m-dimensional Euclidean distance. The separation measure D_i^+ and D_i^- of each alternative or barriers from the PIS and PIN was as given in Table 6.7.

The relative closeness to the ideal solution were calculated, and ranking of the alternatives in descending order was done as in Table 6.8.

From the ranking, it was observed that the significant barriers or challenges obtained were "ISFGA" followed by "IAATP," "financial constraints (FC)," and "URHCW" that requires appropriate considerations and management (Figure 6.3).

6.4 CONCLUSION

The infections that occur from the wastes produced in healthcare sectors are more in comparison to any other types of wastes. An inadequate as well as inappropriate knowledge of bio-medical wastes management among healthcare personals might have serious health-outcomes and an adverse threat to the environment. Bio-medical wastes are generated

TABLE 6.7 Calculating Separation Measures D_i^+ and D_i^-

	B_1	B_2	B_3	B_4	B_5	B_6	B_7	B_8	B_9	B_{10}
D_i^+	0.523682	0.109242	0.075909	0.12829	0.109242	0.109242	0.113788	0.097477	0.094956	0.125032
D_i^-	–0.04921	–0.04012	–0.07345	–0.02107	–0.04012	–0.13485	–0.03557	–0.05188	–0.0544	–0.02433

during treatment, diagnosis, or immunization of human beings or animals or in investigative performances. These are not only hazardous, but also injurious to human beings or animals and harmful to the environment. Thus, effective management of bio-medical wastes becomes a legal as well as social responsibility to keep the living things safe and cleaner environment.

TABLE 6.8 Separation Measures and the Relative Closeness Coefficient

Barriers	Closeness Coefficient	Normalized	Ranking
B_1	–0.10372	0.003472	8
B_2	–0.58042	0.019432	5
B_3	–29.8699	0.999999	1
B_4	–0.19651	0.006579	7
B_5	–0.58042	0.019432	4
B_6	5.265933	–0.1763	10
B_7	–0.45475	0.015225	6
B_8	–1.13779	0.038092	3
B_9	–1.34136	0.044907	2
B_{10}	–0.2416	0.008089	9

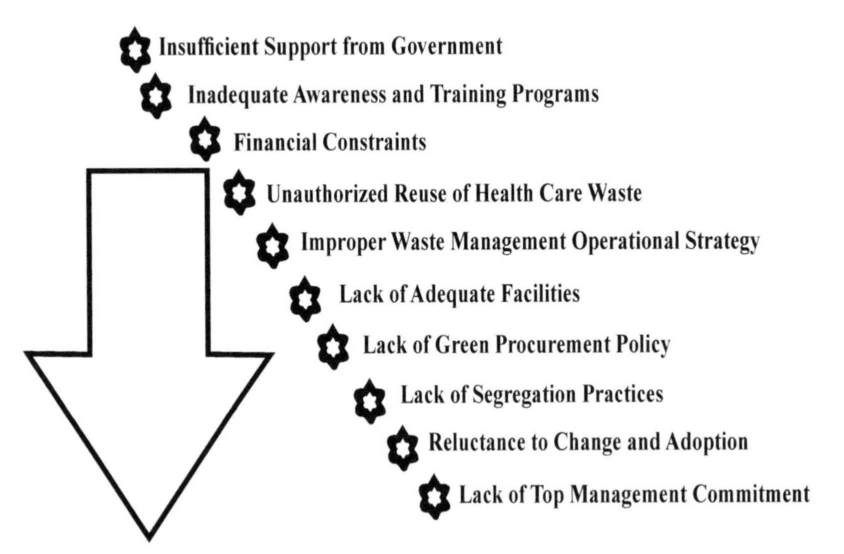

FIGURE 6.3 Challenges and barriers of bio-medical wastes management in descending order.

KEYWORDS

- **bio-medical wastes**
- **effluent treatment plant**
- **environment protection act**
- **hepatitis-B-virus**
- **human immunodeficiency virus**
- **ranking**

REFERENCES

Abah, S. O., & Ohimain, E. I., (2011). Healthcare waste management in Nigeria: A case study. *J. Public Health Epidemiol., 3*, 99–110.

Ajmal, S., & Ajmal, M., (2017). Knowledge and practices of bio-medical waste management among paramedic staff of Jinnah hospital, Lahore. *Biologia., 63,* 59–66.

Akter, N., & Trankler, J., (2003). An analysis of possible scenarios of medical waste management in Bangladesh. *Manag. Environ. Qual., 14*(2), 9.

Ali, S., Mahmood, U., Malik, A. U., Aziz, F., Naghman, R. B., & Ahmed, I., (2015). Current hospital waste management practices in Pakistan: Case study and curative measures. *Public Health Prev. Med., 1,* 125–129.

Amanullah, A. S., & Uddin, J., (2008). Dynamics of health behavior regarding hospital waste management in Dhaka, Bangladesh: A dysfunctional health belief model. *Int. Q. Commun. Health Educ., 29*(4), 363–380.

Asadullah, M. D., Karthik, G. K., & Dharmappa, B., (2013). A study on knowledge, attitude and practices regarding biomedical waste management among nursing staff in private hospitals in Udupi city, Karnataka, India. *International Journal of Geology, Earth and Environmental Sciences, 3*(1), 118–123.

Askarian, M., Vakili, M., & Kabir, G., (2004). Hospital waste management status in university hospitals of the Fars province, Iran. *Int. J. Environ. Health Res., 14,* 295–305.

Askarian, M., Vakili, M., & Kabir, G., (2004). Results of a hospital waste survey in private hospitals in Fars province, Iran. *Waste Manag., 24*(4), 347–352.

Athavale, A. V., & Dhumale, G. B., (2010). A study of hospital waste management at a rural hospital in Maharashtra. *Journal of ISHWM, 9*(1), 21–31.

Battle, L. C., (1994). Regulation of medical waste in the United States. *Pace Environ. Law Rev., 11*(2).

Bhagawati, G., Nandwani, S., & Singhal, S., (2015). Awareness and practices regarding bio-medical waste management among health care workers in a tertiary care hospital in Delhi. *Indian J. Med. Microbiol., 33,* 580–582.

Bhatt, S., Kohli, M., Patel, K., et al., (2013). Evaluation of awareness regarding biomedical waste management in institute of ophthalmology, Ahmedabad, Gujarat. *Natl. J. Integr. Res. Med., 4*(2), 4.

Bio-Medical Waste Management Rules, (2016). Published in the Gazette of India, Extraordinary: Part II, Section 3, Sub-Section (i). *Government of India Ministry of Environment, Forest and Climate Change.* Notification; New Delhi.

Biswas, A., Amanullah, A. S. M., & Santra, S. C., (2011). Medical waste management in the tertiary hospitals of Bangladesh: An empirical enquiry. *ASA Univ. Rev., 5*(2), 10.

Chartier, Y., Emmanuel, J., Pieper, U., Pru¨ss, A., Rushbrook, P., & Stringer, R., (2014). *Safe Management of Wastes from Healthcare Activities* (2nd edn.). World Health Organization (WHO), Geneva, Switzerland.

Das, S. K., & Biswas, R., (2016). Awareness and practice of biomedical waste management among healthcare providers in a tertiary care hospital of West Bengal, India. *Int. J. Med. Public Health, 6,* 19–25.

David, J. J., & Shanbag, P., (2016). Awareness and practices regarding biomedical waste management among healthcare workers in a tertiary care hospital in Delhi: Comment. *Indian J. Med. Microbiol., 34,* 391, 392.

Debalkie, D., & Kumie, A., (2017). Healthcare waste management: The current issue in Menelik II referral hospital, Ethiopia. *Curr. World Environ., 12,* 42–52.

Delmonico, D. V., De G., Santos, H. H. Dos Pinheiro, M. A. P., De Castro, R., & De Souza, R. M., (2018). Waste management barriers in developing country hospitals: Case study and AHP analysis. *Waste Management and Research, 36*(1), 48–58. Retrieved from: https://doi.org/10.1177/0734242X17739972.

Demissie, F., (2014). Hazardous waste management by healthcare institutions, Addis Ababa: Implementation of laws and regulation. *Ethiop. J. Environ. Stud. Manage, 7,* 134–141.

Deress, T., Hassen, F., Adane, K., & Tsegaye, A., (2018). Assessment of knowledge, attitude, and practice about biomedical waste management and associated factors among the healthcare professionals at Debre Markos town healthcare facilities, Northwest Ethiopia. *J. Environ. Public Health,* 1–10.

Directorate General of Health Services (Bangladesh), (2005). *Medical Waste: Risk Assessment, Financial Analysis and Correlates.*

Directorate General of Health Services (Bangladesh), (2012). *Medical Waste Management Training Manual.*

Doylo, T., Alemayehu, T., & Baraki, N., (2018). Knowledge and practice of health workers about healthcare waste management in public health facilities in Eastern Ethiopia. *J. Community Health, 43,* 1–8.

Facility Guidelines Institute, (2010). *Guidelines for Design and Construction of Hospital and Health Care Facilities.* Chicago: American Society for Healthcare Engineering of the American Hospital Association.

Federal Ministry of Health (FMoH), (2008). *Healthcare Waste Management National Guidelines.* Hygiene and Environmental Health Development, FMoH, Addis Ababa, Ethiopia.

Food, Medicine and Healthcare Administration and Control Authority (FMHACA), (2005). *Healthcare Waste Management Directive.* FMHACA, Addis Ababa, Ethiopia.

Government of India, (1989). *Hazardous Wastes (Management and Handling) Rules.* Ministry of Environment and Forests, New Delhi. Available at: http://envfor.nic.in/legis/hsm/hsm1.html (accessed on 18 February 2021).

Government of India, (2000). *Municipal Solid Wastes (Management and Handling) Rules*. Ministry of Environment and Forests, New Delhi. Available at: https://www.mpcb.gov.in/sites/default/files/solid-waste/MSWrules200002032020.pdf (accessed on 18 February 2021).

Gupta, S., & Boojh, R., (2006). Report: Biomedical waste management practices at Balrampur Hospital, Lucknow, India. *Waste Management Research, 24,* 584–591.

Gupta, S., Boojh, R., Mishra, A., & Chandra, H., (2009). Rules and management of biomedical waste at Vivekananda polyclinic: A case study. *Waste Management, 29,* 812–819.

Halbwachs, H., (1994). Solid waste disposal in district health facilities. *World Health Forum, 15*(4), 363–367.

Harhay, M. O., Halpern, S. D., Harhay, J. S., & Olliaro, P. L., (2009). Health care waste management: A neglected and growing public health problem worldwide. *Trop Med. Int. Health, 14,* 1414–1417.

Hassan, A. A., Tudor, T., & Vaccari, M., (2018). Healthcare waste management: A case study from Sudan. *Environments, 5,* 89. doi: 10.3390/environments5080089.

Hassan, M. M., Ahmed, S. A., Rahman, K. A., & Biswas, T. K., (2008). Pattern of medical waste management: existing scenario in Dhaka City, Bangladesh. *BMC Public Health, 8,* 36.

Haylamicheal, I. D., Dalvie, M. A., Yirsaw, B. D., & Zegeye, H. A., (2011). Assessing the management of healthcare waste in Hawassa city, Ethiopia. *Waste Manage. Res., 29,* 854–862.

Holla, R., Darshan, B. B., Sorake, N., Unnikrishnan, B., Thapar, R., Mithra, P., Kumar, N., et al., (2015). Knowledge and practices regarding biomedical waste management among healthcare professionals in tertiary care hospitals of Mangalore, India. *Int. J. Community Med. Public Health, 2*(4), 656–659. doi: http://dx.doi.org/10.18203/23946040.ijcmph20151066.

Hwang, C. L., & Yoon, K. P., (1981). *Multiple Attributes Decision Making Methods and Applications.* Berlin: Springer-Verlag.

IARC, (1985). *Laboratory Decontamination and Destruction of Carcinogens in Laboratory Waste: Some Antineoplastic Agents*. IARC Scientific Publications, No. 73. Lyon: International Agency for Research on Cancer; 1985. Available from: http://www.iarc.fr (accessed on 19 February 2021).

INCLEN Program Evaluation Network (IPEN), (2014). Study Group, New Delhi, India. Bio-medical waste management: Situational analysis and predictors of performances in 25 districts across 20 Indian States. *Indian J. Med. Res., 139,* 141–153.

Ismail, I. M., Kulkarni, A. G., Kamble, S. V., et al., (2013). Knowledge, attitude and practice about bio-medical waste management among personnel of a tertiary health care institute in Dakshina Kannada, Karnataka. *Al Ameen J. Med. Sci., 6*(4), 5.

Kaiser, B., Eagan, P. D., & Shaner, H., (2001). Solutions to health care waste: Life-cycle thinking and "green" purchasing. *Environmental Health Perspectives, 109*(3), 205–207.

Kishore, J., Goel, P., Sagar, B., & Joshi, T. K., (2000). Awareness about biomedical waste management and infection co troll among dentists of a teaching hospital in New Delhi, India. *Indian J. Dent. Res., 11*(4), 157–161.

Krohling, R. A., & Pacheco, A. G. C., (2015). A-TOPSIS: An approach based on TOPSIS for ranking evolutionary algorithms. Information technology and quantitative management (ITQM 2015). *Procedia Computer Science, 55,* 308–317.

Kumari, R., Srivåstava, K., Wakhlu, A., & Singh, A., (2013). Establishing biomedical waste management system in medical university of India-a successful practical approach. *Clinical Epidemiology and Global Health, 1,* 131–136.

Mathur, V., Dwivedi, S., Hassan, M., & Misra, R., (2011). Knowledge, attitude, and practices about biomedical waste management among healthcare personnel: A cross-sectional study. *Indian J. Community Med., 36*(2), 143–145.

Mostafa, G. M., Shazly, M. M., & Sherief, W. I., (2009). Development of a waste management protocol based on assessment of knowledge and practice of healthcare personnel in surgical departments. *Waste Manag., 29*(1), 430–439.

Muduli, K., & Barve, A., (2012a). Challenges to waste management practices in Indian health care sector. *International Conference on Environment Science and Engineering (IPCBEE), 2.* IACSIT Press, Singapore.

Muduli, K., & Barve, A., (2012b). Barriers to green practices in health care waste sector: An Indian perspective. *International Journal of Environmental Science and Development, 3*(4), 393–399.

Nagaraju, B., Padmabathi, G. V., Puranik, D. S., et al., (2013). A study to assess knowledge and practice on bio-medical waste management among the healthcare providers working in PHCs of Bagepalli, Taluk with the view to prepare informational booklet. *Int. J. Med. Biomed. Res., 2*(1), 8.

National Research Council, (1989). *Safe Disposal of Infectious Laboratory Waste.* Biosafety in the Laboratory: Prudent practices for handling and disposal of infectious materials. Washington, DC: The National Academies Press.

Niyongabo, E., Jang, Y. C., Kang, D., & Sung, K., (2019). Current treatment and disposal practices for medical wastes in Bujumbura, Burundi. *Environ. Eng. Res., 24*(2), 211–219. Retrieved from: https://doi.org/10.4491/eer.2018.095 on 20 Jan. 2020.

Ozder, A., Teker, B., Eker, H., et al., (2013). Medical waste management training for healthcare managers: A necessity? *J. Environ. Health Sci. Eng., 11*(1), 1–8.

Pandit, N. B., Mehta, H. K., Kartha, G. P., & Choudhary, S. K., (2005). Management of bio-medical waste: Awareness and practices in a district of Gujarat. *Indian J. Public Health, 49*(4), 245–247.

Patil, A. D., & Shekdar, A. V., (2001). Healthcare waste management in India. *Journal of Environmental Management, 63,* 211–220.

Pennsylvania Department of Environmental Protection (PA-DEP), (2000). 25 PA Code, disposal of infectious and chemotherapeutic waste: Chapter No. 284. *Regulated Medical and Chemotherapeutic Waste: Disposal of Infectious and Chemotherapeutic Waste*. Web link: http://www.pacode.com/secure/data/025/chapter284/chap284toc.html (accessed on 19 February 2021).

Rahman, M., Shahab, S., Malik, R., & Azim, W., (2001). A study of waste generation, collection, and disposal in a tertiary hospital. *Pakistan J. Med. Res., 4*(1), 5.

Rao, P. H., (2008). Report: Hospital waste management-awareness and practices: A study of three states in India. *Waste Manage Res., 26*(3), 297–303.

Rao, S. K. M., Ranyal, R. K., Bhatia, S. S., & Sharma, V. R., (2004). Biomedical waste management: An infrastructural survey of hospitals. *Medical Journal Armed Forces India, 60*(4), 379–382.

Ruoyan, G., Lingzhong, X., Huijuan, L., et al., (2010). Investigation of health care waste management in Binzhou District, China. *Waste Manag., 30*(2), 246–250.

Sachan, R., Patel, M. L., & Nischal, A., (2012). Assessment of the knowledge, attitude and practices regarding bio-medical waste management amongst the medical and paramedical staff in tertiary healthcare center. *Int. J. Sci. Res. Publ., 2*(7), 6.

Saini, S., Nagarajan, S. S., & Sarma, R. K., (2005). Knowledge, attitude and practices of bio-medical waste management amongst staff of a tertiary level hospital in India. *J. Acad. Hosp. Adm., 17*(2), 1–12.

Sari, R. M., Rizkya, I., Syahputri, K., Anizar, & Siregar, I., (2018). Alternative of raw material's suppliers using TOPSIS method in chicken slaughterhouse industry. TALENTA-CEST 2017. *IOP Conf. Series: Materials Science and Engineering, 309,* 012041. doi: 10.1088/1757-899X/309/1/012041.

Sarker, M. A. B., Harun-Or-Rashid, M., Hirosawa, T., Abdul, H. M. S. B., Siddique, M. R. F., Sakamoto, J., & Hamajima, N., (2014). Awareness and barriers towards medical waste management in Bangladesh. *Med. Sci. Monit., 20,* 2590–2597.

Sarotra, P., Medhi, B., Kaushal, V., Kanwar, V., Gupta, Y., & Gupta, A. K., (2016). Health care professional training in biomedical waste management at a tertiary care hospital in India. *J. Biomed. Res., 30,* 168–170.

Secretariat of the Stockholm Convention, (2006). Revised draft guidelines on best available techniques and provisional guidance on best environmental practices of the Stockholm Convention on persistent organic pollutants. Geneva: Secretariat of the Stockholm Convention.

Shinee, E., Gombojav, E., Nishimura, A., et al., (2008). Healthcare waste management in the capital city of Mongolia. *Waste Manag., 28*(2), 435–441.

Singh, A., Kumari, R., Wakhlu, A., Srivastava, K., Wakhlu, A., & Kumar, S., (2014). Assessment of biomedical waste management in a government healthcare setting of North India. *International Journal of Health Sciences and Research, 4*(11), 203–208.

Singh, V. P., Biswas, G., & Sharma, J. J., (2007). Biomedical waste Management: An emerging concern in Indian hospitals. *Indian J. Forensic Med. Toxicol., 1*(2), 6.

The Gazette of India, (1998). *Biomedical Wastes (Management and Handling) Rules*. India: Ministry of Environment and Forests, Government of India.

The Gazette of India, (1998). *Biomedical Waste (Management and Handling) Rules*. Extraordinary, Part II, Section 3, Subsection (ii), India: The Gazette of India, No. 460.

UN, (2009). *UN Recommendations on the Transport of Dangerous Goods* (16th edn.). Geneva: United Nations. Available from: http://www.unece.org/trans/danger/publi/unrec/rev16/16files_e.html (accessed on 19 February 2021).

UN, (2010). *Guidelines for Safe Disposal of Unwanted Pharmaceuticals in and After Emergencies*. Geneva: World Health Organization.

United Nations Industrial Development Organization (UNIDO), (2010). *UNIDO Launches of 40 Million Dollar Project to Help India Dispose of Bio-Medical Waste*. Available from: http://www.unido.org/news/ press/unido-waste-2.html (accessed on 19 February 2021).

Verma, L. K., (2010). Managing hospital waste is difficult: How difficult? *Journal of ISHWM, 9*(1), 46–50.

WHO, (2004b). *Review of Health Impacts from Microbiological Hazards in Healthcare Wastes.* Geneva: World Health Organization.

WHO, (2007). *WHO Core Principles for Achieving Safe and Sustainable Management of Health-Care Waste*. Geneva: World Health Organization. Available from: https://www.who.int/water_sanitation_health/publications/hcwprinciples/en/ (accessed on 18 February 2021).

Windfeld, E. S., & Brooks, M. S. L., (2015). Medical waste management: A review. *Journal of Environmental Management, 163,* 98–108.

Wojciech, S., (2013). *Journal of Theoretical and Applied Computer Science, 7*(3), 40–50.

World Health Organization (WHO), (2011). *Wastes from Health-Care Activities*. Factsheet No. 253. Available from: http://www.who.int/mediacentre/factsheets/fs253/en/ (accessed on 19 February 2021).

World Health Organization Guidance: WHO, (2004a). *Safe Health Care Waste Management: Policy Paper*. Geneva: World Health Organization. Available from: http://www.who.int/water_sanitation_health/medicalwaste/hcwmpolicy/en/index.html (accessed on 19 February 2021).

Yazie, T. D., Tebeje, M. G., & Chufa, K. A., (2019). Healthcare waste management current status and potential challenges in Ethiopia: A systematic review. *BMC Res. Notes, 12,* 285. Retrieved from: https://doi.org/10.1186/s13104-019-4316-y on 26 Dec. 2019.

CHAPTER 7

WASTE MANAGEMENT PRACTICES TO MAINTAIN SUSTAINABLE SUPPLY CHAIN MANAGEMENT: A CASE STUDY ON THE THERMAL POWER SECTOR

ABSTRACT

One of the significant challenges faced all over the world is environmental-pollution and clean energy. Although the thermal power sectors are well known as the highest energy creators, still is getting blamed for environmental pollution creation. So, the thermal industries are in a continuous attempt to resolve their sustainability issues. The power plants in India are not only focusing on sustainable issues but also trying for a sustainable supply chain strategy development to accomplish their operations with due consideration to environmental as well as social issues. In the case of thermal power plants, the "sustainable supply chain management (SSCM)" practices are mostly dependent on the practices such as the utilization of water, ashes, wastes, and energy in addition to proper care of the environment such that environmental, social as well as economic factors remain unaffected. In this chapter, an in-depth focus and analysis were done by considering the SSCM practices of Indian thermal power sectors. Then, the TODIM method was used for prioritizing the waste management practices among coal-based thermal power plants such as privatized, nationalized, state, and imported coal. Simultaneously, their interrelation and correlation were also found.

7.1 INTRODUCTION

Power sectors act as a ladder for human as well as economic progress of any country, which not only improve the persons' worth, but also improve the biotic of this sphere. Electricity consumption acts as measure-indices of a nation's progress level. One the foremost distinguished as well as diversified sector has been recognized as the Indian power sectors within this globe. India is in the race of constructing newer coal-based power plants in addition to the expansion existing ones or enhancing of capacities. A study by "Greenpeace India," the activist group said: "Over Rs. 3,00,000 crore (close to $50 billion) is being wasted on building a further 62 giga-watt of coal power plants, which can remain idle thanks to huge overcapacity within the power sector."

Everywhere throughout the planet, including India, the electricity has been the primary source for all sorts of small scale as well as large scale industries. The Indian power sectors are mainly dependent on coals (e.g., Thermal power plants) to generate 70% of the entire electricity, which may continue or increase in percentage for the subsequent 30 to 40 years based on the predictions. Due to more dependence of the developing countries on thermal power plants for generation of electricity, the entire planet is under the threat of adverse impacts of environmental and social hazards that calls for special attention during this difficulty to help the thermal power industries. One of the major goals of building up the "sustainable development (SD)" practices is to assist the thermal power industries in relieving from environmental effects, expanding the economy of the country and to help in the improvement of societal condition. Completion of industrial operations in a productive manner to ensure in the enhancement of their environmental, economic, and social performances is the aim of sustainable development. By realizing the need of sustainability importance in the supply chain, the development of a replacement-concept, i.e., sustainable supply chain management (SSCM) can be achieved. SSCM has been defined as "the planned, transparent-integration, and realization of an organizations' environmental, social, and economic targets with the systematic dexterity of the key interorganizational business-processes to improve the individuals' economic performances as well as its' supply-chain on long-term basis" (Carter and Rogers, 2008). Already many of the operational thermal power industries in India are involved in SSCM practices with few of them in the urge of implementation like other manufacturing and automobile sectors. Thus, an insight is provided in this research to measure and explore the benefits of SSCM implementation in Indian thermal power industries. Further, a design framework with the suggestion of some important design-requirements for SSCM practices in thermal power sectors of India is provided. With the practice of these design-requirements, the complete SSCM can be implemented properly without any gaps.

The estimated consumption of coals, generation of thermal power and ashes in addition to the NTPC power-plants' in India (Senapati, 2011) are given in Tables 7.1 and 7.2, respectively.

TABLE 7.1 Consumption of Coals, Generation of Thermal Power and Ashes in India

Year	Consumption of Coals in MT	Generation of Thermal Power in MW	Generation of Ashes in MT
1995	200	54,000	75
2000	250	70,000	90
2010	300	98,000	110
2020	350	137,000	140

TABLE 7.2 #NTPC Power-Plants' in India

Power-Station Name (State)	Installed-Capacity in MW
Talcher TPP**, Angul (Odisha)	460
Talcher STPS****, Angul (Odisha)	3000
Vindhyachal STPS****, Sidhi (Uttar Pradesh)	3260
Badarpur TPP**, (NCT-Delhi)	705
Tanda TPP**, Ambedkar Nagar (Uttar Pradesh)	440
Sipat TPP**, Bilaspur (Chhattisgarh)	1000
Feroz Gandhi Unchahar, Raebareli (Uttar Pradesh)	1050
Kahalgaon STPS****, Bhagalpur (Bihar)	2340
Ramagundam STPS****, Karimnagar (Andhra Pradesh)	2600
Farakka STPS****, Murshidabad (West Bengal)	1600
Simhadri STPS****, Visakhapatnam (Andhra Pradesh)	1000
Rihand TPP**, Sonebhadra (Uttar Pradesh)	2000
Korba STPP***, Korba (Chhattisgarh)	2100
Singrauli TPP**, Sonebhadra (Uttar Pradesh)	2000

*NTPC: National Thermal Power-Corporation; **TPP: Thermal power-plant; ***STPP: Super thermal power-plant; ****STPS: Super thermal power station.

The most important problem of thermal power industries are their methods of wastes disposal. The coal reserve of India is about 200 billion tons (bt) and its' annual-production reaches 250 million-tons (mt), approximately. About 70% of this is often utilized in the facility-sectors. In India, unlike most of the developed countries, the ash content within the coal used for generation of power is 30 to 40%. Higher ash-coal means more wear and tear of the plants and machineries, lower thermal-efficiency of the boilers, slogging, choking as well as scaling of the furnace, and most serious of all of them is the outsized amount of ashes generation. India has been ranked fourth within the world in the production of coal-ashes as byproduct-wastes after USSR, the USA, and China, therein order. The management and disposal of ashes may be a major problem in coal-fired thermal power plants. Ash emissions from a spread of coal combustion units show a good range of composition. All elements below number 92 are present in coal-ashes. A 500 megawatt thermal power station releases 200 mt of SO_2, 70 t of NO_2 and 500 t of ashes per day, approximately. The particulates that are considered as a source of pollution, constitute ashes. The fine particles of ash reach the pulmonary region of the lungs and remain there for a longer period of time; they behave like cumulative-poisons. So wastes disposal methods are an important concern for all kind of thermal power plants, which are not only polluting air with the creation of diseases, but also killing the soils fertility at the place of its establishment. Moreover, the fly-ash ponds, contaminated water peripherals are dangerous for all biotic-animals as well as trees. So, thermal power plants are concentrating on clean-technology. All most all kinds of coal suppliers of India like imported, privatized, and nationalized (state-owned) coal mines are trying their best to follow the rules of "Environmental Pollution Control Board" of India. Still, transportation, generation, and waste disposal have been a bigger issue, and a few lacunas are also found in some coal-based thermal power plants. So a study is made in this chapter to explore the waste management processes of thermal power plants.

7.1.1 LITERATURE

The biggest coal-customer has been regarded as the coal power sectors that depend on coals for generation of power. Being a maximum polluting-material, the transportation of coals and their storing is very important everywhere in the planet. Pollutions may be a headache and environmental-condition that need to be a focused issue for discussion. So, a number of efforts are being taken by the government as well as research-units in this issues. It has been found of the media to be always debating and asking for suggestions and newer developments to resolve such problems. As a result, the researchers are at present going on to seek out a better solution by using coals without creating any kind of environmental harmfulness. There is a realization for the current researchers regarding the consideration of environmental as well as societal issues such that greatest revolutions in human-thought can be achieved to unite the whole world in preventing the emissions produced during industrial-activities (Dubey et al., 2015). The last decade in particular has seen an increased pressure to broaden the accountability of the industries beyond economic performances, shareholders to sustainability performances, and for all stakeholders (Labuschagne et al., 2005). Accordingly, an increased interest of SSCM with the triple-bottom-line of sustainability was exhibited by organizations in addressing sustainability in their supply chains (Walker and Jones, 2012). The proper management of raw materials and services from suppliers to manufacturers' or the service-providers to customers' and back with the development of the social and environmental impacts are explicitly considered in SSCM (Grzybowska, 2012). A green or environmental aspect of SSCM is mainly focused on minimizing the adverse environmental consequences of a variety of supply chain activities, where the other social aspect of SSCM ensures ethical and decent working-conditions of different stakeholders, including the suppliers. Through the purchases from local-suppliers, the local financial generation as the economic aspects of SSCM can be ensured (Walker and Jones, 2012). A variety of things influence the implementation of SSCM in industries with no exception for the thermal power industries, which are broadly divided into two categories such as enablers or the factors encouraging SSCM adoptions and barriers hindering SSCM adoptions. The enablers as defined in layman terms' is "an entity that creates it possible or easy." Thus, the enablers are the processes that drive a supply chain to be sustainable (Hussain, 2011). The review of literature reveals that there exists a variety of studies focusing exclusively on the enablers (Diabat et al., 2014; Faisal, 2010; Grzybowska, 2012; Muduli and Barve, 2013; Walker and Jones, 2012) or the barriers (Ageron et al., 2010; Bhattacharyya, 2010; Walker and Jones, 2012). However, this study provides a framework to review both the categories of factors all together, so that it will be easier to spot the relative importance of an enabler with respect to a barrier. This will successfully enable the organizations in identifying their strength levels for handling a specific barrier. Figure 7.1 shows the wastes generation by thermal power plants in India.

FIGURE 7.1 Thermal power plants leading to wastes generation.

The combustion of coal in thermal power plants produces fly-ashes which is a heterogeneous mixture of amorphous as well as crystalline phases, and are usually fine-powdered Ferro-aluminosilicate materials with "Ca, Al, Na, Fe, and Si" as the predominant elements. Some enriched elements found in fly-ash particles are "Mo, B, Se, and S" (Adriano et al., 1980). The particulate devices controlling emissions of fly-ash from the stack into the atmosphere are namely mechanical and electrostatic precipitators and scrubbers (Kumari, 2009). Moreover, the coal-based thermal power generation will continue in playing a dominant role in the future in order to meet the growing energy demand of the country (Sahu et al., 2009). Estimation found the fly-ash generation to increase by 2012 to about 170 million tons, and by 2017 to 225 million-tons (Kumar et al., 2005). In the decreasing order of abundance, the major elements such as "Si, Al, Ca, C, Mg, K, Na, S, Ti, P, and Mn," that exist in the core of the fly-ashes are relatively stable. The primary reason behind this may be because of their un-volatilization during the combustion-stages (EI-Mogazi et al., 1988). A larger quantities of major impurities like "oxides, hydroxides, and sulfates" of calcium and iron, as well as significant quantities of "hazardous-leachable-trace elements" like boron, arsenic, cadmium, manganese, chromium, vanadium, and selenium, are contained in fly-ashes (Querol et al., 1999). Because of the utilization of the organic matters in coal during coal combustion, heat is produced, and thus, the concentrations of trace-elements are increased relative to the source coal (Fernandez et al., 1994). Due to environmental-restrictions, the impurities usually create a negative impact on fly-ash-utilization. The health-hazards as well as environmental-impacts from thermal power plants result in mobilizing toxic and radioactive elements from the residues. The acidic fly-ashes resulting in contaminated-leachates can pose higher toxicity problems for aquatic environment (Roy et al., 1984). The non-toxic soluble elements get dissolved first in weak acids or water (Hulett et al., 1980), while the long-term leaching of toxic-trace elements are associated with slower mobility of elements from glasses, magnetite in addition to the related minerals (Dayan and Paine, 2001). The interaction of surface water as well as groundwater in fly-ashes emplacements take longer time in removing mobile-trace elements from the solid-phase, and on the basis of the 'hydrogeochemical environment' for the use of the fly-ash, the elevated-concentrations are induced over longer time periods by creating potential contamination of associated surface water and groundwater systems. The toxic and mutagenic properties are associated with the fly-ashes containing chromium, that are related to its oxidizing-activities (Dayan and Paine, 2001). Excessive presence of chromium damages the circulatory-systems and results in carcinogenic-changes, due to its mobility in the environment as well as its migrating ability from fly-ash to water-solutions (Soco and Kalembiewicz, 2009). The fly-ashes that contain heavy metals like lead (Pb), are with a significant public-concern owing to its toxicity to human-beings as well as animals. However, lead is extensively used in industrial wastes that contaminate the soil and are widely spread in the environment (Manz, 1999). The associated diseases owing to the presence of heavy-metals in fly-ashes is as shown in Table 7.3.

The "coal combustion residues (CCRs)" are resulted by the coal combustion process. India having miscellaneous quality of coal-reserves contain 30 to 55% ashes. Already the yearly production of CCRs has exceeded 120 MT, and it is expected to be 175 MT by 2012 with the present growth rate of 8 to 10% of power generation (www.tifac.org.in). Only 41% of the generated CCRs in India is being utilized (Kumar and Mathur, 2004). Estimation predicts the coal ash production in India to be about 120 million-tones per annum including both fly-ash and bottom-ash, which is likely to enhance by 2015 A.D to 200 million tons per annum (CEA, 2011). In order to reduce the cost associated with bulk transport of ash laden coals, more emphasis is given in setting-up coal-fired "Super, Ultra, and Mega Power-Projects" on pithead itself, now in India (Palit et al., 1991). A 1000 MW of electricity generating thermal power plant produces about an annual amount of 1.6 million-tons of coal-ash with 80% fly-ashes, and it has been a real challenge for the nation for the management of larger volumes of produced coal-ashes in power plants (Sahu, 2010). The two alternatives left for power plants are either the identification of appropriate ashes-utilization or to dispose them off. Despite of having some beneficial properties by coal-ashes in physical as well as chemical properties, but still it has been a serious concern for regulators, planners, operators, and environmental groups as this is related to health, safety, and environmental risks contaminating air and water qualities (Carloon and Adriano, 1993). Thus, the proper utilization of coal-ashes have drawn considerable attention of technologists, scientists, environmental groups, regulators, and the government. The ash utilization in India has been increased to 50% during 2010–2011 through a number of government and institutional actions, as compared to the early 1990's, when only a very smaller percentage, i.e., 3% of the fly-ashes were used productively with the remaining materials being dumped in vast ash-ponds in slurry-form closer to the power plants (Vibha, 1998).

TABLE 7.3 The Associated Diseases Owing to the Presence of Heavy-Metals in Fly-Ashes

Metals (Symbols)	ppm-Content	Associated Diseases
Cadmium (Cd)	3.4	Hepatic-disorders, Anemia
Nickel (Ni)	77.6	Respiratory-problems, lung-cancers
Arsenic (As)	43.4	Skin-cancers, dermatitis
Antimony (Sb)	4.5	Gastroenteritis
Lead (Pb)	56	Anemia
Chromium (Cr)	136	Cancers

It has been reported of providing 263 billion kWh of electricity in 2013, which was about 90% of whole generated electricity by gas, steam, and combined-cycle thermal power plants (Tavanir, 2014; MOE, 2013). As the combustion of fuels as well as power generation process in plant-units cause a variety pollutions by adversely affecting water, air, and the environment, thus there has been always due attention by energy and environmental experts, in addition to the general-public (Araghi et al., 2012; IL&FS, 2010). However, lesser attention has been provided in the production of semi-solid and solid wastes which are produced directly or indirectly owing to combustions, water-supply systems, and waste-water treatment plants. The quantity as well as quality of solid wastes depend on various factors such as the type of fuel consumption and the type of power plants (WEC, 2008). The solid wastes are produced in more quantity by the steam thermal power plants consuming fuels (mainly oil-furnaces) as well as the electricity generation process (Choban and Winkler, 2010; WB, 1998). The wastes in plant include "sludge, ashes, and health and administrative-trashes." The administrative-trashes occur in all power plants which are recyclable, while the other wastes of industrial-processes like wastewater-treatment are often delivered to contractors. The reuse of plants' waste in industries closer to these plants will reduce pollutants to the environment as well as minimize the waste volumes (Rashidi et al., 2012). For electricity production, two types of process options exist: (a) conventional steam-cycles and (b) combined-cycle power plants. In spite of similarities in both these processes, and additional gas-turbine is inserted in the combined-cycle power plants in front of the steam-cycle. The waste generation amount in both processes may vary depending on differences in processing, process-modifications, and the utilizable type of fuels (Fraas and Ozisik, 1965; Maurstad, 2007). In the case of thermal power plant wastes, the critical issues in management includes the identification of the source-points, qualities, hazardous-contents, and the waste quantities. Moreover, the management strategies identification include efficient waste-collections, appropriate disposal-methods and possible options of recovery/recycling (Woodard, 2001). The widely used coal power plants are producing a larger amount of wastes requiring special attention. Therefore, this requires a strict control measure by the authorities on industrialists in the thermal power sectors to avoid the pollution of environment (Zhang, 2014).

Moreover, one of the promising stage of SSCM implementation is occurring in India, whose adoption by the Indian thermal power sectors has been restricted to only a smaller number of bigger companies. Therefore, an analytical-approach such as AHP has been proposed in this research to take into account various influential factors of SSCM, rather than statistical-approach depending upon larger sample-sizes. A simpler mathematical-method supported by elementary-operations with matrices may be AHP that was developed by Saaty (1970), and it relies on decomposing of the matters into objective, criteria, sub-criteria, alternatives, pair-wise comparison of elements of every level with respect to its' immediate upper-level factors employing a nine-point scale and the generation of priority-vector (Muduli and Barve, 2015). It has the power of accommodating qualitative-attributes in an organized-manner and ranking the impact of the elements on the whole-system (Sambasivan and Fei, 2008), with the target or goal (SSCM implementation) that occupy the highest-level position of the hierarchy, and various criteria and sub-criteria occupy positions within the subsequent utility-level.

7.2 METHODOLOGY

Thermal industries are browsing highly competitive environment for getting profits, fulfilling demands, maximizing benefits, and most vital is to follow guidelines for waste disposal, which can be achieved by the implementation of SSCM in their regular practices. So, in order to operate proper waste disposal methods to achieve the goals of SSCM implementation in thermal power industries, a standard questionnaire was designed with the assistance of educationalist with the industry experts' and through review of literature review that consisted of 66 questionnaires. Then, 100 questionnaires were sent to the executives of different thermal industrial sectors in India by mail, post, and private-contacts. The respondents were advised to reply each item of the questionnaire using a five-point like rt-scale (1 = totally disagree, 2 = partially disagree, 3 = No opinion, 4 = Partially agree, 5 = totally agree). The questionnaire details is given in Appendix 7.1. Among the respondents, 72 responses were obtained, and therefore, the response rate was 70%. Then, the data was fed for statistical analysis like principal component analysis to seek out the items or questionnaires falling under six dimensions SSCM for waste management. Then, the TODIM method was implemented to seek out the best techniques of waste management implemented among privatized, nationalized, or imported coal suppliers to avoid pollution among thermal power plants of India. The steps followed in this work were as illustrated in Figure 7.2.

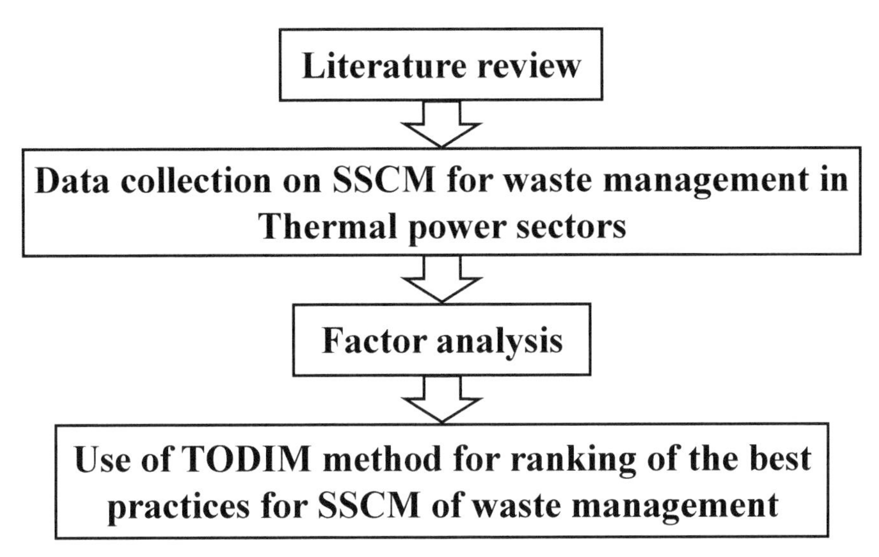

FIGURE 7.2 Steps followed in this work.

7.3 RESULTS AND DISCUSSION

Statistical analysis was performed to the collected data of SSCM in Indian thermal industries like "correlation analysis and Kaiser-Meyer-Olkin (KMO) test." By the use of the SPSS 22.0 version, the correlation analysis on 72 useful responses was performed by the use of principal component method followed by varimax rotation. After the analysis, the total variance explained percentage was found to be 73.5%, which was a suitable value for the principal component varimax rotated factor loading (Johnson and Wichern, 2002). The interior-consistency of this particular survey data was tested by computing the Cronbach's alpha (α). The worth of alpha for every dimension is shown in Tables 7.4. The worth of KMO, which may be a measure of sampling adequacy was found to be 0.505, indicating that the correlation analysis test has correctly proceeded and therefore, the sample used is adequate because of the minimum acceptable value of KMO is 0.5 (Othman and Owen, 2001). Therefore, it is often concluded that the matrix is not suffering from multi co-linearity or singularity. The results of "Bartlett test of sphericity" show that it was highly significant (significance = 0.000), which indicated that the correlation analysis processes was correct and suitable for testing multidimensionality (Othman and Owen, 2001). The factors obtained after correlation analysis were "Waste management, Ash utilization, Biodiversity-conservation, Energy management and Promotion of renewable-energy, Reduction in air-emissions, Water management," respectively.

TABLE 7.4 Factor Analysis

Dominant Factors	Items	Factor-1	Factor-2	Factor-3	Factor-4	Factor-5	Factor-6
Waste management	C1	0.705					
	C2	0.684					
	C9	0.557					
	C10	0.617					
	C11	0.646					
	C12	0.597					
	C13	0.519					
	C14	0.558					
	C15	0.573					
	C16	0.521					
	C20	0.666					
	C21	0.564					
	C22	0.730					
	C24	0.542					
	C26	0.568					
	C27	.630					
	C30	0.615					
	C37	0.505					
	C63	0.512					
Biodiversity-conservation	C33		–0.544				
	C34		–0.551				

TABLE 7.4 *(Continued)*

Dominant Factors	Items	Factor-1	Factor-2	Factor-3	Factor-4	Factor-5	Factor-6
	C36		–0.530				
	C38		–0.629				
	C39		–0.697				
	C40		–0.551				
	C46		0.616				
	C47		0.613				
	C48		0.621				
	C49		0.638				
Reduction in air-emissions	C18			0.591			
	C32			0.548			
	C41			0.633			
	C43			0.825			
	C44			0.682			
	C45			0.512			
Ash utilization	C4				0.763		
	C5				0.744		
	C7				0.593		
Energy management and promotion of renewable-energy	C51					–0.758	
	C52					–0.658	
	C53					–0.595	
	C54					–0.550	
Water management	C29						0.503
	C55						0.513
	C57						0.513
	C59						0.523
	C66						0.566

In Table 7.5, the 47 items were explained under six factors (Waste management, Ash utilization, Biodiversity-conservation, Energy management and promotion of renewable-energy, Reduction in air-emissions,

Water management) which clarifies the extent of sustainability measures in thermal power industries.

Table 7.6 shows the correlation of dimensions of SSCM that explain wastes disposal methods. Pearson-coefficient of correlation (r) is employed to explain the strength and direction of the connection between two variables. The computation of the Pearson-coefficient of correlation is performed to get an understanding of the connection between all variables within this study. The typical score of the multiple items for a dimension is computed as each dimension of the questionnaire is measured by multiple items. Strong correlation is found between "Waste management, Biodiversity-conversion, and Reduction of air-emissions." Ash utilization shows poor relation with "Energy management and promotion of renewable energy.

TABLE 7.5 Naming of the Constructs

C1	Mine fills
C2	Fill-materials in cement
C9	Partial replacement of lime-aggregates in concrete-works
C10	Roads and railways embankments
C11	Land filling-materials
C12	Manufacturing of ash-bricks
C13	Utilization in agriculture depending on crops
C14	Land development
C15	Installation of units based on ash-brick product making
C16	Installation of bio-methanation plants
C20	Green-belts, afforestation, and energy-plantation (plantation of trees in around and townships)
C21	Reclamation of abandoned ash-pond
C22	Ecological monitoring and scientific studies
C24	Socio-economic studies
C26	Geo-hydrological studies
C27	Use of waste products and services
C30	Capacity addition in old-plants within existing-land
C37	Dust-extraction systems and dust-suppression systems
C63	Rain-water harvesting

TABLE 7.5 *(Continued)*

C33	Fabric-filters and dust-collectors
C34	Flue-gas stacks
C36	Neutralization-pits
C38	Installation of cooling towers
C39	Ash-dykes and ash-disposal system
C40	Dry-ash extraction system
C46	Gas-conditioning system
C47	Monitoring systems to know the efficacy of used systems
C48	Monitoring of environmental-parameters
C49	Environment-reviews
C18	Installation of vermin-composting for canteen-wastes
C32	Use of particulate-scrubbers
C41	Ultrasonic dust-suppression systems
C43	Clean-coal technology
C44	Process of desulfurization
C45	Process of denitrification
C4	Light-weight aggregates
C5	Partial cement replacement
C7	Grouting-materials
C51	Installation of roof-top "Solar PV Plant"
C52	Installation of solar water heater
C53	Installation of Magna-drive with crusher-motor and conveyor in CHP
C54	Implementation of variable frequency drive in NOx injection-pump and raw water pump
C29	Reduction in land-requirements for main plant and ash disposal-areas in newer units
C55	Replacement of "sodium vapor lamps" with CFL/LED lamps
C57	Efficient use of fuels (coal, natural gas, and fuel-oil)
C59	Reduction in water-requirement for main plant and ash disposal-areas through recycle and reuse of water
C66	Attention paid to prevent water pollution at water catchment areas, in particular for drinking-water, mineral-water, and aquatic-life

TABLE 7.6 Correlation Analysis

	Waste Management	**Biodiversity-Conservation**	**Reduction in Air-Emissions**	**Ash Utilization**	**Energy Management and Promotion of Renewable-Energy**	**Water Management**
Waste management	1.000					
Biodiversity-conservation	0.351	1.000				
	P = 0.003					
Reduction in air-emissions	0.343	0.400	1.000			
	P = 0.004	P = 0.001				
Ash utilization	–0.067	–0.033	–0.074	1.000		
	P = 0.584	P = 0.791	P = 0.546			
Energy management and promotion of renewable-energy	–0.083	–0.333	–0.346	–0.159	1.000	
	P = 0.500	P = 0.005	P = 0.004	P = 0.192		
Water management	–0.313	–0.007	0.031	0.245	–0.116	1.000
	P = 0.009	P = 0.955	P = 0.800	P = 0.003	P = 0.344	

Table.7.7 shows the items regression analysis for the selected 47 items.

TABLE 7.7 Regression Analysis of Items

Variable	Adjusted Total Mean	Total Standard Deviation	Item-Adjusted Total Correlation	Multiple Correlation	Cronbach's Alpha
C1	162.4	20.01	0.584	1.000	0.858
C2	162.5	20.00	0.672	1.000	0.857
C3	162.7	20.04	0.508	1.000	0.859
C4	162.6	20.25	0.399	1.000	0.862
C5	162.6	19.96	0.630	1.000	0.857
C6	162.5	19.88	0.662	1.000	0.857
C7	162.7	20.37	0.323	1.000	0.863
C8	162.7	20.22	0.434	1.000	0.861
C9	162.3	20.16	0.480	1.000	0.860
C10	162.5	20.17	0.451	1.000	0.861
C11	162.3	20.26	0.411	1.000	0.862
C12	162.4	20.00	0.585	1.000	0.858
C13	162.4	20.11	0.495	1.000	0.860
C14	162.5	20.24	0.450	1.000	0.861
C15	162.4	20.17	0.530	1.000	0.860
C16	162.2	20.12	0.544	1.000	0.859
C17	162.5	19.99	0.561	1.000	0.858
C18	162.30	20.19	0.562	1.000	0.860
C19	162.4	20.31	0.299	1.000	0.864
C20	162.2	20.37	0.379	1.000	0.863
C21	162.3	20.51	0.179	1.000	0.866
C22	162.4	20.35	0.282	1.000	0.864
C23	162.2	20.25	0.456	1.000	0.861
C24	162.6	20.22	0.451	1.000	0.861
C25	162.4	20.50	0.159	1.000	0.866
C26	162.4	20.51	0.142	1.000	0.867
C27	162.3	20.06	0.521	1.000	0.859
C28	162.6	20.24	0.360	1.000	0.863
C29	162.51	20.28	0.398	1.000	0.862
C30	162.9	20.37	0.268	1.000	0.864

TABLE 7.7 *(Continued)*

Variable	Adjusted Total Mean	Total Standard Deviation	Item-Adjusted Total Correlation	Multiple Correlation	Cronbach's Alpha
C31	162.3	20.52	0.120	1.000	0.867
C32	162.7	20.60	0.077	1.000	0.868
C33	162.7	20.36	0.291	1.000	0.864
C34	162.7	20.40	0.243	1.000	0.865
C35	162.7	20.41	0.212	1.000	0.865
C36	163.1	20.31	0.306	1.000	0.864
C37	162.7	20.40	0.234	1.000	0.865
C38	162.5	20.20	0.404	1.000	0.862
C39	162.9	20.83	0.107	1.000	0.872
C40	163.0	20.65	0.026	1.000	0.869
C41	162.9	20.80	0.092	1.000	0.870
C42	162.9	20.90	0.175	1.000	0.872
C43	162.3	20.11	0.510	1.000	0.860
C44	162.6	20.84	0.115	1.000	0.872
C45	162.7	20.56	0.120	1.000	0.867
C46	162.4	20.26	0.410	1.000	0.862
C47	162.4	20.57	0.093	1.000	0.868

Then, the TODIM method was implemented to prioritize waste management practices among coal (privatized, nationalized, state, and imported coal). The criteria are taken as waste management practices in a thermal power plant. TODIM (an acronym in Portuguese for Interactive and Multi-Criteria Decision Making) may be a multi-criteria method introduced by Gomes and Lima (1992a, b). This value function, which shows an S-shaped growth curve, allows the behavior of the choice maker to be reflected with reference to gains and losses. The TODIM method is used for prediction on a pairwise-comparison of the alternatives, and calculates for every criterion on the dominance of one alternative over another by the use of worth-function introduced by Kahneman and Tversky (1979) (see also Tversky and Kahneman, 1992). The TODIM method makes use of a global-measure notion for computable values by the application of the paradigm of "Prospect Theory". This method is based on a description already-proved by empirical-evidence of effective decision-making by people in risky circumstances (Gomes et al., 2009; Gomes and Rangel, 2009) as illustrated in Table 7.8.

The traditional TODIM method decision-making steps can be summarized as follows:

1. Normalize η=[dij]m×n η = [d i j] m × n into η′=[d′ij]m×n η ′ = [d i j ′] m × n.

TABLE 7.8 Matrix of Partial-Desirability

Alternatives	Criteria					
	C1	**C2**	**…**	**Cj**	**…**	**Cm**
A1	P11	P12	…	P1j	…	P1m
A2	P21	P22	…	P2j	…	P2m
…	…	…	…	…	…	…
Ai	Pi1	Pi2	…	Pij	…	Pim
…	…	…	…	…	…	…
An	Pn1	Pn2	…	Pnj	…	Pnm

2. Calculating the dominance degree of ηi over each alternative ηt based on cj.
3. Computing the overall value of δ(ηi) δ (η i) with formula:

$$\Phi_c(A_i, Aj) = \begin{cases} \sqrt{\dfrac{w_{rc}(P_{ic} - P_{jc})}{\sum_{c=1}^{m} w}} & \text{if } (P_{ic} - P_{jc}) > 0, \quad (2) \\ 0 & \text{if } (P_{ic} - P_{jc}) > 0, \quad (3) \\ \dfrac{-1}{\theta}\sqrt{\dfrac{\left(\sum_{c=1}^{m} w_{rc}\right)(P_{jc} - P_{ic})}{w_{rc}}} & \text{if } (P_{ic} - P_{jc}) > 0, \quad (4) \end{cases}$$

- $\delta(A_i, A_j)$ represents the measurement of dominance of alternative *i* over alternative *j*;
- *m* is the number of criteria;
- *c* is any criterion, for $c = 1., m$;
- $-w_{rc}$ is the substitution rate of the criterion *c* by the reference criterion *r*;

- P_{ic} and P_{jc} are, respectively the performances of the alternatives i and j in relation to c;
- θ is the attenuation factor of losses.

4. Choosing the best alternative by rank values of δ(ηi), the alternative with maximum value is the best choice (Tables 7.9–7.13).

$$\xi_i = \frac{\sum_{j=1}^{n} \delta(A_i, A_j) - \min \sum_{j=1}^{n} \delta(A_i, A_j)}{\max \sum_{j=1}^{n} \delta(A_i, A_j) - \min \sum_{j=1}^{n} \delta(A_i, A_j)}$$

TABLE 7.9 Kind of Criteria

kind of Criteria	Waste Management	Biodiversity-Conservation	Reduction of Air-Emissions	Ash Utilization	Energy Management and Promotion of Renewable-Energy	Water Management
	–1	1	–1	1	1	1

TABLE 7.10 Data for Criteria

Weights of criteria	0.25	0.25	0.25	0.25	0.25	0.25
Kind of criteria	–1	1	–1	1	1	1
	C1	**C2**	**C3**	**C4**	**C5**	**C6**
A1	3	4	5	5	4	5
A2	4	4	4	4	4	5
A3	5	5	1	2	2	2
A4	5	5	5	1	1	1
A5	5	5	5	4	4	5
A6	5	5	5	1	1	5
A7	4	2	4	3	4	1
A8	5	4	5	4	5	4
A9	5	4	4	4	5	4
A10	4	4	4	4	4	3
A11	1	2	2	1	3	4
A12	2	2	2	2	3	5
A13	4	3	4	2	3	4
A14	5	5	5	5	5	5
A15	4	4	1	3	4	4

TABLE 7.10 *(Continued)*

Weights of criteria	0.25	0.25	0.25	0.25	0.25	0.25
A16	4	5	4	2	4	2
A17	4	4	3	4	5	4
A18	4	4	4	5	4	4
A19	5	5	4	4	5	5
A20	2	4	5	4	4	4
A21	5	4	1	4	5	4
A22	4	4	4	4	5	5
A23	2	5	5	5	4	4
A24	4	5	4	4	5	5
A25	5	4	5	4	5	4
A26	5	2	2	2	4	5
A27	4	4	5	4	2	5
A28	2	1	2	1	2	1
A29	2	5	4	5	4	4
A30	5	4	5	2	1	2
	4	5	5	5	5	4
SUM	5	4	2	1	4	5
	5	5	4	2	1	4

TABLE 7.11 Decision-Matrix of Case-Study

	C1	C2	C3	C4	C5	C6
A1	0.075	1	0.083	5.000	1	1
A2	0.056	1	0.016	4.000	1	1
A3	0.037	1.250	0.083	2.000	0.500	0.400
A4	0.094	1.250	0.033	1	0.250	0.200
A5	0.018	1.250	0.067	4.000	1	1
A6	0.075	1.250	0.016	1	0.250	1
A7	0.094	0.500	0.083	3.000	1	0.200
A8	0.094	1	0.033	4.000	1.250	0.800
A9	0.094	1	0.067	4.000	1.250	0.800
A10	0.075	1	0.083	4.000	1	0.600
A11	0.018	0.500	0.033	1	0.750	0.800
A12	0.037	0.500	0.083	2.000	0.750	1
A13	0.0377	0.7500	0.0833	2.0000	0.7500	0.8000
A14	0.018	1.250	0.083	5.000	1.250	1

TABLE 7.11 *(Continued)*

	C1	C2	C3	C4	C5	C6
A15	0.094	1	0.067	3.000	1	0.800
A16	0.075	1.250	0.083	2.000	1	0.400
A17	0	1	0	4.000	1.250	0.800
A18	0	1	0	5.000	1	0.800
A19	0	1.250	0	4.000	1.250	1
A20	0	1	0	4.000	1	0.800
A21	0	1	0	4.000	1.250	0.800
A22	0	1	0	4.000	1.250	1
A23	0	1.250	0	5.000	1	0.800
A24	0	1.250	0	4.000	1.250	1
A25	0	1	0	4.000	1.250	0.800
A26	0	0.500	0	2.000	1	1
A27	0	1	0	4.000	0.500	1
A28	0	0.250	0	1	0.500	0.200
A29	0	1.250	0	5.000	1	0.800
A30	0	1	0	2.000	0.250	0.400

TABLE 7.12 Normalized Decision-Matrix

	A1	A2	A3	A4	SUM
A1	0	0.569	0.167	0.065	0.801
A2	–3.418	0	–0.618	–0.591	–4.628
A3	–8.143	–7.120	0	0.303	–14.96
A4	–9.498	–8.218	–5.220	0	–22.93

TABLE 7.13 Final-Ranking

Alternatives	Global-Dominance [G(ai)]	Relative Overall-Value [V(ai)]		Ranking
A1	0.801	1	1	1
A2	–4.628	0.771	0.771	2
A3	–14.96	0.336	0.336	3
A4	–22.93	0	0	4

7.4 CONCLUSION

The implementation of SSCM may be a comprehensive and difficult phenomenon, but the demand and requirement for a pollution-free environment as well as clean energy is increasing everywhere in the planet. As because the thermal power sectors are blamed for creating environmental pollution, so they are more focused on sustainability issues and also, subsequently trying in the development of sustainable supply chain strategies to carry-out their operations while respecting the social as well as environmental issues. From this study, it was found that nationalized suppliers are ranked first for following best waste-disposal guidelines. This study will provide an insight to privatized companies of thermal power sectors to follow the best waste-disposal guidelines. In this chapter, factor-analysis shows the essential-dimensions of supplier-selection criteria for thermal power industries as "fuel-quality, fuel-cost, credibility of suppliers, and sustainability." After finding the criteria and sub-criteria of supplier-selection to be prioritized, the alternatives for AHP was applied on various alternatives of Indian thermal power plants like "privatized coal suppliers, state or nationalized suppliers, and imported coal suppliers." Among these suppliers, it was found that "state or nationalized suppliers" were ranked first, followed by "imported suppliers." It means these suppliers are following rules of environmental protection more as compared to "privatized suppliers." So to stay on the competitive market, the "privatized suppliers" need to give more emphasis on the most vital criteria like "sustainability," than on "fuel cost and quality." In the future, more power sectors must be involved to see the certainty of the result. Some other sorts of deciding methods are often implemented during this study for justification of the content.

KEYWORDS

- **factor analysis**
- **sustainable supply chain management**
- **thermal power plant**
- **TODIM**
- **transparent-integration**
- **waste management**

REFERENCES

Adriano, D. C., Page, A. L., Elseewi, A. A., Chang, A. C., & Straughan, I., (1980). Utilization and disposal of fly-ash and other coal residues in terrestrial ecosystems: A review. *Journal of Environmental Quality, 9,* 333–344.

Ageron, B., Gunasekaran, A., & Spalanzani, A., (2012). Sustainable supply management: An empirical study. *Int. J. Production Economics, 140,* 168–182.

Anwar, R. M., Masud, A. K. M., Abedin, F., & Hossain, M. E., (2010). QFD for utility services: A case study of electricity distribution company DESCO. *Int. J. Quality and Innovation, 1*(2), 184–195.

Araghi, M. K., Sharzehi, G. H., & Barkhordari, S., (2012). *Journal of Ecology, 38,* 93.

Bhattacharyya, S. C., (2010). Shaping a sustainable energy future for India: Management challenges. *Energy Policy, 38,* 4173–4185.

Carloon, C. L., & Adriano, D. C., (1993). Environmental impacts of coal combustion residues. *Journal of Environmental Quality, 22,* 227–247.

Carter, C. R., & Rogers, D. S., (2008). A framework of sustainable supply chain management: Moving toward new theory. *International Journal of Physical Distribution and Logistics Management,* 38(ED-5), 360–387.

Central Electricity Authority (CEA), (2011). *Report on Fly Ash Generation at Coal/Lignite Based Thermal Power Stations and its Utilization in the Country for the Year "2010–11.* New Delhi.

Choban, A., & Winkler, I., (2010). Analysis of influence of the thermal power plant wastewaters discharge on natural water objects. Water treatment technologies for the removal of high-toxicity pollutants. *NATO Science for Peace and Security Series C: Environmental Security,* 225–234.

Dayan, A., & Paine, A., (2001). Mechanisms of chromium toxicity, carcinogenicity and allergenicity. *Journal of Human Toxicology, 20,* 439–510.

Delgado, J., Saraiva, P. A., & Almeida, A. T. D., (2007). Electrical power delivery improvement in Portugal through quality function deployment. *Ninth International Conference, Electrical Power Quality and Utilization* (pp. 1–6). doi: 10.1109/EPQU.2007.4424223.

Diabat, A., Kannan, D., & Mathiyazhagan, K., (2014). Analysis of enablers for implementation of sustainable supply chain management: A textile case. *Journal of Cleaner Production.* doi: 10.1016/j.jclepro.2014.06.081.

Dubey, R., Gunasekaran, A., Wamba, S. F., & Bag, S., (2015). Building theory of green supply chain management using total interpretive structural modeling. *IFAC-Papers Online, 48*(3), 1688–1694.

EI-Mogazi, E., Lisk, D., & Weinstein, L., (1988). A review of physical, chemical, and biological properties of fly-ash and effects on agricultural ecosystems. *Journal of the Science of the Total Environment, 74,* 1–37.

Faisal, M. N., (2010). Sustainable supply chains: A study of interaction among the enablers. *Business Process Management Journal, 16*(3), 508–529.

Fernandez, T. J. L., De Carvalho, W., Cabanas, M., Querol, X., & Lopez-Soler, (1994). A mobility of heavy metals from coal fly-ash. *Journal of Environmental Geology, 23,* 264–270.

Fraas, A. P., & Ozisik, M. N. A., (1965). *Comparison of Gas-Turbine and Steam-Turbine Power Plants for Use with All-Ceramic Gas-Cooled Reactors.* Oak Ridge, Tennessee.

Gomes, L., & Lima, M. M. P. P., (1992a). From modeling individual preferences to multicriteria ranking of discrete alternatives: A look at prospect theory and the additive difference model. *Foundations of Computing and Decision Sciences, 17*(3), 171–184.

Gomes, L., & Lima, M. M. P. P., (1992b). TODIM: Basics and application to multicriteria ranking of projects with environmental impacts. *Foundations of Computing and Decision Sciences, 16*(4), 113–127.

Gomes, L., & Rangel, L. A. D., (2009). An application of the TODIM method to the multicriteria rental evaluation of residential properties. *European Journal of Operational Research, 193*, 204–211.

Gomes, L., Rangel, L. A. D., & Maranhao, F. J. C., (2009). Multicriteria analysis of natural gas destination in Brazil: An application of the TODIM method. *Mathematical and Computer Modeling, 50,* 92–100.

Handfield, R., Walton, S. V., & Sroufe, R., (2005). Integrating environmental management and supply chain strategies. *Business Strategy and the Environment, 14*(1), 1–19.

Handfield, R., Walton, S. V., Sroufe, R., & Melnyk, S. A., (2002). Applying environmental criteria to supplier assessment: A study in the application of the analytical hierarchy process. *European Journal of Operational Research, 141*(1), 70–87.

Hulett, L. D., Weinberger, A. J., Northcutt, K. J., & Ferguson, M., (1980). Chemical species in fly-ash from coal-burning power plants. *Journal of Science, 210,* 1356–1364.

IL&FS, (2010). *Technical EIA Guidance Manual for Thermal Power Plants.* IL&FS EcoSmart Limited, Hyderabad, India.

Johnson, R. A., & Wichern, D. W., (2002). *Applied Multivariate Statistical Analysis.* Prentice-Hall, New Jersey, Upper Saddle River, NJ.

Kahneman, D., & Tversky, A., (1979). Prospect theory: An analysis of decision under risk. *Econometrica, 47,* 263–292.

Katarzyna, G., (2012). Sustainability in the supply chain: Analyzing the enablers. In: Golinska, P., & Romano, C. A., (eds.), *Environmental Issues in Supply Chain Management, Eco Production* (pp. 25–40). Springer-Verlag, Berlin Heidelberg.

Keeney, R. L., & Raiffa, H., (1993). *Decisions with Multiple Objectives: Preferences and Value Tradeoffs.* Cambridge: Cambridge University Press.

Kumar, V., & Mathur, M., (2004). Fly ash-gaining acceptance as building. *Proceedings of the Seminar on Recent Trends in Building Materials* (pp. 55–65). Bhopal, India.

Kumar, V., Mathur, M., Sinha, S. S., & Dhatrak, S., (2005). *Fly-Ash Environmental Savior.* New Delhi, India: Report submit to fly-ash utilization program (FAUP) to TIFAC, DST.

Kumari, V., (2009). Physicochemical properties of fly-ash from thermal power station and effects on vegetation. *Global Journal of Environmental Research, 3*(2), 102–105.

Labuschagne, C., Brent, A. C., & Claasen, S. J., (2005a). Environmental and social impact considerations for sustainable project life cycle management in the process industry. *Corporate Social-Responsibility and Environmental Management, 12*(1), 38–54.

Labuschagne, C., Brent, A. C., & Van, E. R. P. G., (2005). Assessing the sustainability performance of industries. *Journal of Cleaner Production, 13*(4), 373–385.

Lewis, W. G., Pun, K. F., & Lalla, T. R. M., (2006). Empirical investigation of the hard and soft criteria of TQM in ISO 9001 certified small and medium-sized enterprises. *International Journal of Quality and Reliability Management, 23*(8), 964–985.

Manz, O. E., (1999). Coal fly-ash: A retrospective and future look. *Journal of Fuel, 78*(2), 133–136.

Maurstad, (2007). *An Overview of Coal based Integrated Gasification Combined Cycle (IGCC) Technology*. Cambridge, MA. Retrieved from: https://sequestration.mit.edu/pdf/LFEE_2005-002_WP.pdf (accessed on 19 February 2021).

Mittal, V., Ross, W. T., & Baldasare, P. M., (1998). The asymmetric impact of negative and positive attribute-level on overall satisfaction and repurchase intentions. *Journal of Marketing, 62*(3), 33–48.

MOE, (2013). *Energy balance 1391*. Planning office, Department of Electrical Energy, Ministry of Energy.

Mohammed, H., (2011). *Modeling the Enablers and Alternatives for Sustainable Supply Chain Management*. Concordia Institute for Information Systems Engineering, Canada, PhD Thesis.

Nordmann, L. H., & Luxhoj, J. T., (2000). Neural network forecasting of service problems for aircraft structural component grouping. *Journal of Aircraft, 37*(2), 332–338.

Palit, A., Gopal, R., Dubey, S. K., & Mondal, P. K., (1991). Characterization and utilization of coal ash in the context of super thermal power stations. *Proceedings International Conference on Environmental Impact of Coal Utilization* (pp. 154, 155). IIT, Bombay.

Querol, X., Juan, C. U., Andres, A., Carles, A., Angel, L. S., & Felicia, P., (1999). Extraction of major soluble impurities from fly-ash in open and closed leaching systems. *International Ash Utilization Symposium, Center for Applied Energy Research.* University of Kentucky.

Rashidi, Z., Karbassi, A. R., Ataei, A., Ifaei, P., Samiee-Zafarghandi, R., & Mohammadi-zadeh, M. J., (2012). *Int. J. Environ. Res., 6,* 875.

Roy, W. R., Griffin, R. A., Dickerson, D. R., & Schuller, R. M., (1984). Illinois basin coal fly-ashes 1: Chemical characterization and solubility. *Journal of Environmental Science Technology, 18,* 734–745.

Sahu, P., (2010). *Characterization of Coal Combustion by-Products (CCBS) for Their Effective Management and Utilization*. Thesis.

Sahu, S. K., Bhangare, R. C., Ajmal, P. Y., Sharma, S., Pandit, G. G., & Puranil, V. D., (2009). Characterization and quantification of persistent organic pollutants in fly-ash from coal-fueled thermal power stations in India. *Journal of Microchemical, 92,* 92–96.

Satapathy, S., (2014). ANN, QFD, and ISM approach for framing electricity utility service in India for consumer satisfaction. *Int. J. Services and Operations Management, 18*(4), 404–428.

Satapathy, S., Patel, S. K., & Mishra, P. D., (2012). Discriminate analysis and neural network approach in water utility service. *Int. J. Services and Operations Management, 12*(4), 468–489.

Satapathy, S., Patel, S. K., Mahapatra, S. S., & Biswas, A., (2013). A framework for electricity utility service by QFD technique. *Int. J. Services Sciences, 5*(1), 58–73.

Senapati, M. R., (2011). Fly ash from thermal power plants-waste management and overview. *Current Science, 100*(12), 1791–1794.

Soco, E., & Kalembiewicz, J., (2009). Investigations on chromium mobility from coal fly-ash. *Journal of Fuel, 88,* 1513–1519.

Tavanir, (2014). *Detailed Stats of Iran's Electricity Industry for Managers at 2013*. Company of production, distribution and transfer of electric power of Iran.

Vibha, K., (1998). *Environmental Assessment of Fly Ash in its Disposal Environment at F.C.I, Sindri, Jharkhand (India).* MTech Thesis, Submitted to Indian School of Mines Dhanbad.

Walker, H., & Jones, N., (2012). Sustainable supply chain management across the UK private sector. *Supply Chain Management: An International Journal, 17*(1), 15–28.
WB, (1998). Thermal power: Guidelines for new plants. *Pollution Prevention and Abatement Handbook*. World Bank Group.
WEC, (2008). *Implementation of an Effective Waste Management System into the Thermal Power Plants*. WEC Regional Energy Forum.
Woodard, F., (2001). *Industrial Waste Treatment Handbook* (Vol. 1542). CEUR Workshop Proceedings.
www.tifac.org.in (accessed on 19 February 2021).
Zhang, X., (2014). *Management of Coal*. IEA Clean Coal Center. Retrieved from: http://bookshop.iea-coal.org.uk/report/80575//83258/Management-of-coal-combustion-wastes-CCC-231 (accessed on 19 February 2021).

APPENDIX 7.1 Questionnaire for SSCM Measures in Thermal Power Industries

SL. No.	SSCM Measures	1	2	3	4	5
1.	Mine fills					
2.	Fill-materials in cement					
3.	Building blocks					
4.	Light-weight aggregates					
5.	Partial cement replacement					
6.	Road sub-base					
7.	Grouting-materials					
8.	Filler in asphalt-mix for roads					
9.	Partial replacement of lime-aggregates in concrete-works					
10.	Roads and railways embankments					
11.	Land filling-materials					
12.	Manufacturing of ash-bricks					
13.	Utilization in agriculture depending on crops					
14.	Land-development					
15.	Installation of units based on ash-brick product making					
16.	Installation of bio-methanation plants					
17.	Installation of facility for conversion of domestic wastes to organic-fertilizers					
18.	Installation of vermin-composting for canteen-wastes					
19.	Segregation, storages, and disposals of hazardous and non-hazardous wastes into separate pits for healthy-sanitation of township					
20.	Green-belts, afforestation, and energy-plantation (Plantation of trees in around and townships)					

APPENDIX 7.1 *(Continued)*

SL. No.	SSCM Measures	1	2	3	4	5
21.	Reclamation of abandoned ash-pond					
22.	Ecological monitoring and scientific studies					
23.	Environment impact-assessment studies					
24.	Socio-economic studies					
25.	Ecological monitoring program					
26.	Geo-hydrological studies					
27.	Use of waste products and services					
28.	Advanced or eco-friendly technologies like (IGCC)					
29.	Reduction in land-requirements for main plant and ash disposal-areas in newer units					
30.	Capacity addition in old-plants within existing-land					
31.	Installation of "electrostatic precipitator"					
32.	Use of particulate-scrubbers					
33.	Fabric-filters and dust-collectors					
34.	Flue-gas stacks					
35.	Use of "Low-NOX Burners"					
36.	Neutralization-pits					
37.	Dust-extraction systems and dust-suppression systems					
38.	Installation of cooling towers					
39.	Ash-dykes and ash-disposal system					
40.	Dry-ash extraction system					
41.	Ultrasonic dust-suppression systems					
42.	"Zero-emission" technology					
43.	Clean-coal technology					
44.	Process of de-sulfurization					
45.	Process of de-nitrification					
46.	Gas-conditioning system					
47.	Monitoring systems to know the efficacy of used systems					
48.	Monitoring of environmental-parameters					
49.	Environment-reviews					
50.	Upgradation and retrofitting of pollution control system					
51.	Installation of roof-top "Solar PV Plant"					
52.	Installation of solar water heater					

APPENDIX 7.1 *(Continued)*

SL. No.	SSCM Measures	1	2	3	4	5
53.	Installation of Magna-drive with crusher-motor and conveyor in CHP					
54.	Implementation of variable frequency drive in NOx injection-pump and raw water pump					
55.	Replacement of "sodium vapor lamps" with CFL/LED lamps					
56.	Reduction in fuel requirement through more efficient combustion and adoption of state-of-the-art technologies such as "supercritical boilers"					
57.	Efficient use of fuels (coal, natural gas, and fuel-oil)					
58.	Reduction in water requirement for main plant and ash disposal areas through recycle and reuse of water					
59.	Reduction in water-requirement for main plant and ash disposal-areas through recycle and reuse of water					
60.	Liquid-waste treatment plants and management system					
61.	Sewage-treatment plants and facilities					
62.	Coal-setting pits/oil-setting pits					
63.	Rain-water harvesting					
64.	Protection of ground-water from pollution					
65.	Prevention of water pollution at water catchment areas, particularly for drinking-water, mineral-water, and aquatic-life					
66.	Attention paid to prevent water pollution at water catchment areas, in particular for drinking-water, mineral-water, and aquatic-life					

CHAPTER 8

CONCLUSION

8.1 PRESENT SCENARIO OF WASTE MANAGEMENT

In the Indian context, the waste management situation is different from that in developed countries. The different factors that play a significant role in influencing waste management in various regions in the country are poor organizational structures, lack of technical equipment as well as laws and finance, etc. The existing waste management system is not effective for ensuring nationwide disposals as the main-forms of waste disposals in many developing countries. Normally, in many regions, the wastes are collected un-controllably, transported, ill or un-treated, and landfilled, which creates a high risk for public health and damages the natural resources by contaminating water, air, and soil.

This book demonstrates different novel solutions for waste management in developing countries, particularly in India. Although many technical processes related to waste treatment are transferred from industrialized-nations, but due to the lack of specialist-knowledges, the expensive technical-processes that are not meeting the local requirements, cannot be used in a successful manner. Thus, increasing awareness about the environment and escalating problem pressures are encouraging the policy-makers for designing an orderly and effective waste management system. This book accomplished its' objectives by providing landscapes and by conducting a critical analysis of the applications in MCDM techniques for proper waste management. It may be noted that these MCDM methods are at present in a wide range of usage in different decision-making applications and also can be more effectively used for the waste management context.

Based on the studies made in the preceding chapters, the weaknesses of the current waste management system in the Indian context can be summarized as follows:

- Although in many countries there are laws related to waste management, but it could not be implemented successfully because of lacking of necessary infrastructures as well as institutions, such as treatment facilities and laboratories.
- The waste management services in many countries are not properly covered by an efficient financial system. As the waste fees are very low that covers a lower percentage of the necessary costs in order to achieve a sustainable waste management.
- There is a lack of support for the specialized departments in the relevant ministries as well as offices.
- There is a lack of the necessary knowledge and specialized personnel.
- In many cities, the waste management structures are lacking in quality management system tools, and also there is no good control for the evaluation of the realized activities.

Therefore, in order to develop and improve the present situations of the waste management system, many regions in the country must necessarily develop the necessary advisory structure which is possible only with the support of decision-makers. In addition, there must be an international-trend toward the cooperation and integration with the private-sectors and to get involved in management services.

8.2 EVOLUTION OF ZERO WASTE (ZW) CONCEPT

In order to satisfy the ever-growing consumption cultures, the extraction of resources and production of goods has constantly been expanded (Lopez, 1994). Different ranges of consumer-products, like clothes, white goods as well as electronic-products are now used as everyday-goods, once treated as luxury-items (Crocker, 2013). The production process has transformed into a complex-system utilizing composites and hazardous-materials. Because of the generation of large amount of wastes from mixed sources by damaging the environment and also, expensive for sustainably management, the decision-makers find no other option except choosing an inefficient and environmentally-polluting solutions of waste management such as landfill. The shortages of landfill-sites in urban-areas puts the waste authorities for looking for an alternative system of waste management (Wen et al., 2009).

In recent decades, a visionary waste management system for waste-problems such as "Zero Waste (ZW)" has been suggested as an alternative solution (Connett, 2013a). ZW has become As an aspirational ambition to tackle the waste problems, a number of cities such as Vancouver, San Francisco, and Adelaide have adopted the ZW objectives as strategies for waste management (Connett, 2006; SF-Environment, 2013). However, the perceptions and applications of the zero-waste concept is taken in different ways by different professionals in waste management systems. For instance, despite of prohibition in incineration and landfills in the ZW concepts, but a number of studies have still claimed of achieving ZW objectives through the utilization of waste-to-energy techniques, such as incineration, as a part of the waste treatment process (ZWIA, 2009). Over time, a number of innovations have been implemented for waste management such as landfills, composting, recycling, in addition to complex treatment methods. But, in order to achieve a true sense of sustainability in waste management systems, the ZW has been considered as the most holistic-innovation of the 21st century (Zaman and Lehmann, 2011). In 1973, Palmer (2004) has initially used the "ZW" term in order to recover the resources from chemicals, and since the late 1990s, this concept has obtained much public attention. The concept of ZW has been worldwide adopted by a number of organizations with a primary target of ZW disposals to landfills. Canberra has been regarded as the first city in the world in adopting an official ZW objective (Connett, 2013b; Snow and Dickinson, 2003). In New Zealand, the "ZW New Zealand Trust" was established in 1997 to support the waste minimization initiation through ZW movements. The ultimate goal targeted by the Trust was to create "a closed-loop materials-economy for making reusable, repairable, and recyclable products, and an economy that minimizes as well as eliminates waste ultimately" (Tennant-Wood, 2003). The first comprehensive ZW plan in the United States was undertaken by Del Norte County, California in 2000, and later on, the "California Integrated Waste Management Board" adopted the ZW targets as strategic waste management plans, in 2001 (Connett, 2013b). The major events/milestones for the zero-waste target development is as shown in Table 8.1.

8.3 ENHANCEMENT OF WASTE MANAGEMENT STRATEGIES THROUGH INTERNET OF THINGS (IOT)

With the rapid growth of cities, the amount of waste generation in the urban areas is also increasing proportionately. Thus, these wastes must be

TABLE 8.1 Major Events/Milestones for the Zero Waste Target Development (Connett, 2013b)

Country (Year)	Events/Milestones
USA (1970s)	Paul Palmer first coined the term of "zero waste."
USA (1986)	The formation of "National-Coalition against Mass-Burn Incineration" was taken place.
USA (1988)	The "Pay-As-You-Throw (PAYT)" system was introduced by Seattle.
USA (1989)	The "California Integrated-Waste-Management Act" was passed in order to attain 25% waste-diversion from landfills by 1995, and to 50% by 2000.
Sweden (1990)	The "Extended-Producer-Responsibility" was introduced by Thomas Lindqvist.
Australia (1995)	The "No Waste by 2010" bill was passed by Canberra.
New Zealand (1997)	The establishment of "Zero Waste New Zealand Trust." Conference was organized by the "California Resource-Recovery-Association (CRRA)" on 'zero waste.'
USA (1998)	Inclusion of "Zero waste" as guiding principles in "North Carolina, Seattle, Washington, along with Washington, DC."
USA (1999)	Organization of "zero waste" conference by the CRAA in 'San Francisco.'
USA (2000)	The formation of "Global-Alliance for Incinerator-Alternatives."
USA (2001)	Publication of "A Citizen's Agenda for Zero Waste" by grass-roots recycling-network.
New Zealand (2002)	Publication of the book "Cradle-to-Cradle." Establishment of "Zero Waste International Alliance." Holding of the First "ZW Summit" in 'New Zealand.'
USA, Australia (2004)	A working definition of "zero waste" was given by 'ZWIA.' Adoption of "ZW business principles" by 'GRRN.' Establishment of "Zero Waste SA" in 'South Australia.'
USA (2008)	Adoption of a "zero waste producer-responsibility-policy" by the "Sierra Club."
USA (2012)	At the "Cannes Film Festival," the documentary film Trashed was premiered. Establishment of the "Zero Waste Business Council" in the USA.

managed appropriately so as to ensure sustainable development along with a decent standard of living for all residents. The management of wastes is a complex issue concerned to the constrained regions for resources. The municipal solid waste (MSW) management from conventional approach suffers from lack of sustainable collections (Al Mamun et al., 2016), also includes inadequate data on collection-time as well as locations, in-efficient monitoring-systems of status of waste-bins, inappropriate waste collection-transportation systems, and collection delays. Thus, a creative, reliable approach is needed on solid waste management, in terms of design of local models with good involvement of all stakeholders (Wilson et al., 2012). For improving the waste data collection quality, the use of technology can be adopted to make the cities into smart cities. A Smart City refers to "Smart Economy-Mobility-Environment-People-Living-Governance, that are built through the smart-combination of independent. Self-decisive and awareness of concerned citizens endowments as well as activities (Balandin et al., 2015). Therefore, for achieving green-city-environment and well-being of the citizens, a smart city process of waste collection is a primary requirement and thus, there should consideration of its quality. Moreover, this can be achieved through the application of "information and communications technologies (ICT)" solutions through "internet of things (IoT)" (Balandin et al., 2015), which allows things and people to be connected anytime as well as anywhere with anyone and anything by the use of any path or network (Kumar, 2014). The IoT concept enables An easy access as well as interaction with different kinds of devices such as home-appliances, surveillance-cameras, actuators, monitoring-sensors, displays, vehicles, etc., are possible with the IoT concept (Akpakwu et al., 2017; Abu-Mahfouz and Hancke, 2018; Zanella et al., 2014). The IoT concept was developed in parallel to "wireless sensor networks (WSNs)," which is an important aspect in IoT and these enables to monitor the changes as well as advances in environmental aspects and also increasingly adopted for a variety of applications (Mesmoudi et al., 2013). Moreover, the "multiagent system (MAS)" which is a set of software-agents interacted for solving problems beyond the knowledge and capacities of individuals (Potiron et al., 2013). Different researches have been done by the use of agents to provide simulated-solutions for different smart cities, for instance intelligent-waste-collection (Karadimas et al., 2006), and generic-models for smart-cities (Longo et al., 2014). El-Mougy et al. (2015) have proposed Garbage Management model using IoT for Smart-Cities in organizing the garbage collection systems of commercial as well as residential areas, for which the waste material levels in the garbage bin will be detected by the help of ultrasonic-sensors that will be communicated continuously to the authorized control-room through GSM-module. Similarly, Monroy et al. (2014) have proposed a low-cost embedded device for tracking the garbage bins levels by providing a unique ID to each dust-bin in the city for easier identification of filled garbage bins. There project module was divided into two-parts like "transmitter section and receiver section." The transmitter section was composed of "8051 microcontrollers, RF transmitter and sensors," attached to the dustbins; while the sensor was used in detecting the full or empty levels in the dustbins. Furthermore, in order to solve shortest-path problems, the "Geographical information system and Dijkstra algorithm" was used (Sanjeevi and Shahabudeen, 2016), where the traveled distance for waste collection was reduced by 9.93% of the thirteen identified wards, with a reduction of the time as well as associated costs.

8.4 RECYCLING OF WASTE

8.4.1 PLASTIC RECYCLING

Recycling is a simple process to protect our surroundings and make sure the well-being of our community for generations to return. However, the success of recycling depends on the active participation of each member of the community. By participating, people are going to be reducing the quantity of trash that is disposed within the landfill, encouraging the reuse of materials made up of recycled products and continuing the recycling circle. Plastic recycling is imperative in ensuring that natural resources are returned to nature to make sure their sustainability. Plastic was alleged to be the wonder product of the 20th century, but the toxic industrial waste created by it has been dangerous. Therefore, it has become imperative that we recycle all plastic waste.

Plastic recycling is that the process of recovering differing types of plastic material so as to reprocess them into varied other products, unlike their original form. An item made out of plastic is recycled into a special product, which usually cannot be recycled again.

Before any plastic waste is recycled, it must undergo five different stages in order that it are often further used for creating various sorts of products:

1. **Sorting:** It is necessary that each plastic item is separated consistent with its make and sort in order that it are often processed accordingly within the shredding machine.
2. **Washing:** Once the sorting has been done, the plastic waste must be washed properly to get rid of impurities like labels and adhesives. This enhances the standard of the finished product.
3. **Shredding:** After washing, the plastic waste is loaded into different conveyor belts that run the waste through the various shredders. These shredders shred the plastic into small pellets, preparing them for recycling into other products.
4. **Identification and Classification of Plastic:** After shredding, a correct testing of the plastic pellets is conducted so as to determine their quality and sophistication.
5. **Extruding:** This involves melting the shredded plastic in order that it are often extruded into pellets, which are then used for creating differing types of plastic products.

➢ **Benefits.**
- **Conservation of Energy and Natural Resources:** The recycling of plastic helps save tons of energy and natural resources as these are the most ingredients required for creating virgin plastic. Saving petroleum, water, and other natural resources help conserve the balance in nature.
- **Clears Landfill Space:** Plastic waste is accumulated ashore that ought to be used for other purposes. The sole way this plastic waste are often faraway from these areas is by recycling it. Also, various experiments have proven that when another waste is thrown on an equivalent ground as plastic waste, it decomposes faster and emits hazardous toxic fumes after a particular period. These fumes are extremely harmful to the encompassing area as they will cause differing types of lung and skin diseases.

8.4.2 RECYCLING OF BUILDING MATERIAL

Recycling of construction waste is a method to counter risk to construction wastes. So, the invention of proper technology to recycle these materials is of great importance. As an example, concrete waste are often crushed and used as recycled aggregate. Responsible management of waste is an important aspect of sustainable building. During this context, managing waste means eliminating waste where possible; minimizing waste where feasible; and reusing materials which could otherwise become waste. Solid waste management practices have identified the reduction, recycling, and reuse of wastes as essential for sustainable management of resources.

Most construction and demolition waste currently generated within the U.S. is lawfully destined for disposal in landfills regulated under code of federal regulations (CFR) 40, subtitles D and C. In some areas, all or a part of construction and demolition waste stream is unlawfully deposited ashore, or in natural drainages including water, contrary to regulations to guard human health, commerce, and therefore the environment. Businesses and citizens of the U.S. legally eliminate many plenty of building-related waste in solid waste landfills annually. Increasingly, significant volumes of construction-related waste are far away from the waste stream through a process called diversion. Diverted materials are sorted for subsequent recycling, and in some cases reused. Volumes of building-related waste generated are significantly influenced by macroeconomic conditions affecting construction, societal consumption trends, and natural and anthropogenic hazards. In recent years, housing industry awareness of disposal and reuse issues has been recognized to scale back volumes of construction and demolition waste disposed in landfills. Many opportunities exist for the beneficial reduction and recovery of materials that might rather be destined for disposal as waste. Housing industry professionals and building owners can educate and be educated about issues like beneficial reuse, effective strategies for identification and separation of wastes, and economically viable means of promoting environmentally and socially appropriate means of reducing total waste disposed. Organizations and governments can assume stewardship responsibilities for the orderly, reasonable, and effective disposal of building-related waste, promotion of public and industry awareness of disposal issues, and providing stable business-friendly environments for collecting, processing, and repurposing of wastes.

Effective management of building-related waste requires coordinated action of governmental, business, and professional groups and their activities. Several non-governmental organizations and societies within the US promote coordinated action, and have identified best management practices within the interest of public health and welfare (see resources.) Absent coordinated regulations, realistic business opportunities, and therefore the commitment of design and construction professionals and their clients for continual improvement of industry practices, consistent, and stable markets for recovered materials cannot be achieved or sustained.

Management of building-related waste is dear and sometimes presents unintended consequences. However, sense suggests that failure to scale back, reuse, and recycle societal wastes is unsustainable. It stands to reason that efficient and effective elimination and minimization of waste, and reuse of materials are essential aspects of design and construction activity. Creativity, persistence, knowledge of obtainable markets and businesses, and understanding of applicable regulations are important skills for design and construction professionals.

Effective management of building-related waste requires coordinated action of governmental, business, and professional groups and their activities. Several non-governmental organizations and societies within the US promote coordinated action, and have identified best management practices within the interest of public health and welfare (see resources.) Absent coordinated regulations, realistic business opportunities, and therefore the commitment of design and construction professionals and their clients for continual improvement of industry practices, consistent, and stable markets for recovered materials cannot be achieved or sustained.

Management of building-related waste is dear and sometimes presents unintended consequences. However, sense suggests that failure to scale back, reuse, and recycle societal wastes is unsustainable. It stands to reason that efficient and effective elimination and minimization of waste, and reuse of materials are essential aspects of design and construction activity. Creativity, persistence, knowledge of obtainable markets and businesses, and understanding of applicable regulations are important skills for design and construction professionals.

Industrial hygienists perform waste characterization studies and identify components that present known risks to human health and therefore the environment. Specialty contractors provide comprehensive services for identification, verification, removal, handling, and disposal of known and suspect hazardous and dangerous materials in accordance with applicable regulations.

The ability to reuse and recycle materials salvaged from demolition and building sites for reuse and recycling depends on few points like local recycling facilities, market demand, quality, and condition of materials and components, time available for salvage, and emphasis placed on reuse and recycling.

Materials which will generally be recycled from construction sites include:

- Steel from reinforcing, wire, containers, and so on.
- Concrete, which may be weakened and recycled as base course in driveways and footpaths, aluminum, Plastics-grade 1 (PET) and a couple of (HDPE).
- Paper and cardboard, untreated timber, which may be used as firewood or mulched topsoil and paint. A variety of manufacturers/ retailers take back unwanted paint and paint containers.
- Reuse/recycling from deconstruction/demolition sites.

Materials which will generally be recycled from deconstruction/demolition sites includes are siteworks and vegetation-asphalt paving, chain link fencing, timber fencing, trees, concrete-in place and precast concrete, masonry concrete blocks and ornamental concrete, paving stones, bricks, metals reinforcing steel (rebar), steel, steel roofing including flashings and spouting, zinc roofing, interior metal wall studs, cast iron, aluminum, copper including flashings, spouting, claddings, and pipework, lead, electrical, plumbing fixtures timber-hardwood flooring, laminated beams, truss joists, treated, and untreated timbers/posts, joinery, untreated timber generally, engineered timber panels, terracotta tiles, electrical wiring, wool carpet, plastics-grade 1 (PET) and a couple of (HDPE).

Components which will readily be reused includes stairs, timber hardwood flooring, weatherboards, laminated beams, truss joists, treated, and untreated framing, timbers/posts, New Zealand native timber components, thermal insulation fiberglass, wool, and polyester insulation, polystyrene sheets, carpet and carpet tiles, plumbing fixtures like baths, sinks, toilets, taps, service equipment, predicament heaters, electrical fittings like light fittings, switches, thermostats, linings and finishing-architraves, skirtings, wood paneling, specialty wood fittings, joinery, doors, and windows like metal and timber doors, mechanical closures, panic hardware, aluminum windows, steel windows, sealed glass units, unframed glass mirrors, storefronts, skylights, glass from windows and doors, timber, and metal from frames as clay and concrete roof tiles, metal wall and roof claddings, PVC and metal spouting.

Hazardous materials must be disposed of appropriately. Check the wants for removal and disposal of hazardous waste for your local area. As fluorescent light ballasts manufactured before 1978 contain PCBs, fluorescent lamps contain mercury, refrigeration, and air-con equipment-contain refrigerants made using CFCs, batteries that contain lead, mercury, and acid, roof, and wall claddings, pipe insulation, some vinyl flooring, textured ceilings and roofing membrane sheets containing asbestos fibers. Asbestos can cause very serious health problems, and its removal is controlled by law. Lead or materials that contain lead like flashings, paint, bath, and basin wastes.

When cleaning up, materials like cement, sand, paint, and other liquids and solvents, must not be released into the stormwater or sewerage disposal systems. This could be included within the demolition specification.

8.4.3 INDUSTRIAL WASTE RECYCLING

The major generators of commercial solid wastes are the thermal power plants producing coal ash, the integrated Iron and Steel mills producing furnace slag and steel melting slag, non-ferrous industries like aluminum, zinc, and copper producing red mud and tailings, sugar industries generating press mud, pulp, and paper industries producing lime and fertilizer and allied industries producing gypsum.

Coal Ash generally, a 1,000 MW station using coal of three, 500 kilocalories per kg and ash content within the range of 40–50% would wish about 500 hectares for disposal of ash for about 30 years' operation. It is, therefore, necessary that ash should be utilized wherever possible to attenuate environmental degradation. The thermal power station should take under consideration the capital and operation/maintenance cost of ash disposal system also because the associated environmental protection cost, vis-a-vis dry system of collection and its utilization by the thermal power station or other industry, in evaluating the feasibility of such system. The research and development (R&D) administered in India for utilization of ash for creating building materials has proved that ash are often successfully utilized for production of bricks, cement, and other building materials. Indigenous technology for construction of building materials utilizing ash is out there and are being practiced during a few industries. However, large scale utilization is yet to require off. Albeit the complete potential of ash utilization through manufacture of ash bricks and blocks is explored, the number of ash produced by the thermal power plants are so huge that a major portion of it will still remain unutilized. Hence, there is a requirement to evolve strategies and plans for safe and environmentally sound method of disposal.

Integrated Iron and steel mill slag the furnace (BF) and steel melting shop (SMS) slags in integrated iron and steel plants are at the present dumped within the surrounding areas of the steel plants making hillocks encroaching on the agricultural land. Although the BF slag has potential for conversion into granulated slag, which may be a useful staple in cement manufacturing, it is yet to be practiced in a big way. Even the utilization of slag as road subgrade or landfilling is additionally very limited. Phosphogypsum is that the waste generated from the orthophosphoric acid, ammonium phosphate and acid plants. This is often very useful as an artifact. At the present little or no attention has been paid to its utilization in making cement, plasterboard, partition panel, ceiling tiles, artificial marble, fiberboards, etc. Red mud as solid waste is generated in non-ferrous metal extraction industries like aluminum and copper. The red mud at the present is disposed in tailing ponds for settling, which more often than not finds its course into the rivers, especially during monsoon. However, red mud has recently been successfully tried, and a plant has been found out within the country for creating corrugated sheets. Demand for such sheet should be popularized and encouraged to be used. This might replace asbestos which is imported and also banned in developed countries for its hazardous effect. Attempts also are made to manufacture polymer and natural fibers composite panel doors from red mud. Lime sludge, also referred to as lime mud, is generated in pulp and paper mills which is not recovered for reclamation of quicklime to be used except within the large mills. The lime mud disposal by dumping into low-lying areas or into watercourses directly or as run-off during monsoon is not only creating serious pollution problem but also wasting the precious non-renewable resources. The explanations for not reclaiming the quicklime within the sludge after recalcination is that it contains a high amount of silica. Although a couple of technologies are developed to desilicate black liquor before burning, none of the mills within the country are adopting desilication technology. Potential Reuse of Solid Wastes R&D studies conducted by the R&D Institutions like Central Building Research Institute, Roorkee (CBRI) and therefore the National Council for Building Research (NCBR), Ballabgarh reveal that the aforesaid solid wastes features an excellent potential to be utilized within the manufacture of varied building materials (Table 8.2).

TABLE 8.2 Industrial Waste Recycling

Types of Waste	Reuse/Recycle
1. Textile industry (wastewater)	Recycle in process house
2. Alcohol industry	Reuse of energy in process house
3. Food processing	Recycling for irrigation/process house and reuse of energy
4. Viscose rayon	Recovery and reuse of zinc. Foreign exchange saving.
5. Cement industry	Recovery and reuse of cement and clinker dust

Source: Industrial Waste Management, NWMC (1990).

8.4.4 INDUSTRIAL WASTE AND AREAS OF APPLICATION

Fly-ash is employed in cement, staple in Ordinary Portland Cement (OPC), manufacture of oilwell cement, for making sintered fly-ash light-weight aggregates, cement/silicate bonded fly-ash/clay binding bricks and insulating bricks, cellular concrete bricks and blocks, lime, and cement ash concrete. Precast fly-ash in concrete building units, structural fill for roads, construction on sites, land reclamation, etc. It is used as filler in mines, in bituminous concrete, as plasticizer, and as water reducer in concrete and sulfate resisting concrete. It is also used as an amendment and stabilization of soil. Red mud is employed as a corrective material, as a binder, for making construction blocks and as a cellular concrete additive. It is also used for colored composition for concrete and for creating heavy clay products and red mud bricks. It is also utilized in making floor, all tiles and red mud polymer door.

8.4.5 AGRICULTURAL WASTE RECYCLING

Recycling of fruits and agricultural wastes have been one of the foremost as well as important means of utilization in a number of innovative ways, which can help in yielding new products in addition to meeting the wants of essential products that are required in human, animal, and plant nutrition. The pharmaceutical industries also find these wastes to be useful. Microbial technology is out there for recycling and processing of fruit and vegetables waste and following products are often made out of the various processes.

An enormous amount of agricultural wastes are generated at every stage of the food supply chain like harvesting, transportation, storage, marketing, and processing. Thanks to their nature and composition, they deteriorate easily and cause foul smell. The vegetable wastes generated within the markets disposed-off in municipal landfill or dumping grounds. These wastes are often reused for productive purpose like compost preparation, biogas generation, electricity production, etc. The waste from fruit and vegetable processing industries are often used for production of biogas. Biogas is produced by anaerobic digestion of fruit and vegetable wastes. The conversion of fruit and Vegetable Wastes to biogas using anaerobic digestion process may be a viable and commercial option. Bioconversion processes are suitable for wastes containing moisture content above 50% than the thermo-conversion processes. Vegetable wastes, thanks to their high biodegradability nature and high moisture content (75–90%) are an honest substrate for bio-energy recovery through the aerobic digestion process. The wastes generated within the markets are disposed-off in municipal landfill or dumping grounds. These wastes are often reused for productive purpose like electricity production. Collaboration with Atomic research station and thermal power plants are often effective in convert these wastes into a valuable source of energy. Bio-Solutions India, a Hamburg-based company with operations in Ramanagara, near Bengaluru, buys agricultural waste from farmers and makes biodegradable packaging and tableware from it. The patented technology converts agro-waste into self-binding fibers by simply churning them with water in huge machines. The top products are sustainable packaging and tableware that takes only three months to biodegrade. Kriya Labs provides an answer to convert per annum 15–20 million tons of crop stubble that mainly comprises of rice straw which does not have any market as compared to other sorts of agro-waste like wheat straw and bagasse.

They convert rice straw into pulp using their own process. This pulp is often used as an intermediary product for industries like paper (and its derivatives), bioethanol (biofuels in general), fabrics, and other specialty chemicals like cellulose ester, cellulose. The method is environment friendly. The standard if pulp depends on its end application and may be altered accordingly. Farm2Energy, a startup from Punjab founded by Sukhbir Singh Dhaliwal, helps farmers by baling the stubble from fields freed from cost. Founded in July 2016, the firm provides biomass supply solutions to a variety of clients from farmers to landowners and biomass users. The corporate processes paddy straw, corn stover, sugar cane trash and wheat straw and supplies it to the biofuels, biopower, and bio-based industries through integrated supply systems. Additionally, it also helps farmers in managing and utilizing biomass.

8.5 RECYCLING DONE IN OTHER COUNTRIES

Sweden is already known as one of the world's best recycling nations and now it has added a feather in its cap by opening an upcycling shopping mall called ReTuna. ReTuna works on the formula of environment friendly products and sells only repaired or upcycled products. The staff repairs the items dropped by people for restoring and then sell it in the mall. The mall has created 50 jobs and contains a training college for studying recycling. It also has a restaurant, conference facilities and a retail space for local entrepreneurs and artisans.

A Scotland-based company called MacRebur came up with a way to create roads using recycled plastic which are stronger and durable than roads built using asphalt. Asphalt-based roads crack during rains and do not last long. Plastic roads are 60% stronger and last around 10 times longer. Along with this, it also solves the problem of growing plastic pollution. The roads are made with a material that MacRebur calls MR6. The MR6 gets mixed with other construction materials and replaces the need for a wasteful material called bitumen.

The Belgium-based startup has taken the plastic management game a step higher by printing 3D sunglasses out of recycled plastic. The plastic waste is fed into the 3-D printer, melted to form thin strands of plastic wire and then used for the construction of frames. Later, frames are fitted with Italian lenses. The company also urges people to return the glasses once these are used so that it can be recycled into a new pair of glasses. Not only this, to make people use these glasses, the company has come up with five unique designs and three different colors of lenses.

People dump bottles anywhere, and when it rains, all the bottles end up in trenches leading to flood. Mirro wanted a solution to this problem, and then he came up with this idea. Construction of houses starts with the process of filling plastic bottles with soil. This not only reuses plastic bottles but also saves the environment. These houses are cost-effective too, as the materials used are readily available. Mirro is spreading his idea to refugee camps and also teaching marginalized youth how to use bottles to build houses.

KEYWORDS

- **code of federal regulations**
- **information and communications technologies**
- **internet of things**
- **multiagent system**
- **recycling**
- **wireless sensor networks**

REFERENCES

Abu-Mahfouz, A. M., & Hancke, G. P., (2018). Localized information fusion techniques for location discovery in wireless sensor networks. *International Journal of Sensor Networks (IJSNET), 26*(1), 12–25. doi: 10.1504/IJSNET.2017.10007406.

Akpakwu, G. A., Silva, B. J., Hancke, G. P., & Abu-Mahfouz, A. M., (2017). A survey on 5G networks for the internet of things: Communication technologies and challenges. *IEEE Access*, *5*(12), 1–29. doi: 10.1109/ACCESS.2017.2779844.

Al Mamun, M. A., Hannan, A., Hussain, & Basri, H., (2016). Theoretical model and implementation of a real time intelligent bin status monitoring system using rule-based decision algorithms. *Expert Syst. Appl., 48,* 76–88.

Balandin, S., Andreev, S., & Koucheryavy, Y., (2015). *Internet of Things, Smart Spaces, and Next Generation Networks and Systems* (Vol. 9247). Cham: Springer International Publishing.

Connett, P., (2006). Zero waste wins. *Altern. J., 32*(1), 14, 15.

Connett, P., (2013a). Zero waste 2020: Sustainability in our hand. In: Lehmann, S., & Crocker, R., (eds.), *Motivating Change: Sustainable Design and Behavior in the Built Environment*. Earthscan Publication, London.

Connett, P., (2013b). *The Zero Waste Solution*. Chelsea Green Publishing, Vermont.

Crocker, R., (2013). From access to excess: Consumerism, 'compulsory' consumption and behavior change. In: Lehmann, S., & Crocker, R., (eds.), *Motivating Change: Sustainable Design and Behavior in the Built Environment*. Earthscan Publication, London.

El-Mougy, A., Ibnkahla, M., & Hegazy, L., (2015). Software-defined wireless network architectures for the internet of things. In: *40th Annual IEEE Conference on Local Computer Networks* (pp. 804–811).

Karadimas, N. V., Rigopoulos, G., & Bardis, N., (2006). Coupling multiagent simulation and GIS: An application in waste management. *WSEAS Transactions on Systems, 5*(10), 2367–2371.

Kumar, S., (2014). Ubiquitous smart home system using android application. *Int. J. Comput. Networks Commun., 6*(1), 33–43.

Longo, M., Roscia, M., & Lazaroiu, G. C., (2014). Innovating multiagent systems applied to smart city. *Res. J. Appl. Sci. Eng. Technol., 7*(20), 4296–4302.

Lopez, R., (1994). The environment as a factor of production: The effects of economic growth and trade liberalization. *J. Environ. Econ. Manag., 27*(2), 163–184.

Mesmoudi, A., Feham, M., & Labraoui, N., (2013). Wireless sensor networks localization algorithms: A comprehensive survey. *Int. J. Comput. Networks Commun., 5*(6), 45–64.

Monroy, J. G., Gonzalez, J., & Sanchez, C., (2014). Monitoring household garbage odors in urban areas through distribution maps. *Sensors IEEE, 2014.*

Palmer, P. (2004). The Road to Zero Waste: Strategies for Sustainable Communities. Getting to Zero Waste. California: Purple Sky Press.

Potiron, K., El Fallah, S. A., & Taillibert, P., (2013). *From Fault Classification to Fault Tolerance for Multi-Agent Systems.* London: Springer London.

Sanjeevi, V., & Shahabudeen, P., (2016). Optimal routing for efficient municipal solid waste transportation by using ArcGIS application in Chennai, India. *Waste Manag. Res., 34*(1), 11–21.

SF-Environment, (2013). *San Francisco Zero Waste Legislation.* Available from: http://www.sfenvironment.org/zero-waste/overview/legislation (accessed on 18 February 2021).

Tennant-Wood, R., (2003). Going for zero: A comparative critical analysis of zero waste events in southern New South Wales. *Australas. J. Environ. Manag., 10*(1), 46–55.

Wen, L. C., Lin, H., & Lee, S. C., (2009). Review of recycling performance indicators: A study on collection rate in Taiwan. *Waste Manag., 29*(8), 2248–2256.

Wilson, D. C., Rodic, L., Scheinberg, A., Velis, C. A., & Alabaster, G., (2012). Comparative analysis of solid waste management in 20 cities. *Waste Manag. Res., 30*(3), 237–254.

Zaman, A. U., & Lehmann, S., (2011). Challenges and opportunities in transforming a city into a "zero-waste city." *Challenges, 2*(4), 73–93.

Zanella, A., Bui, N., Castellani, A., Vangelista, L., & Zorzi, M., (2014). Internet of things for smart cities. *IEEE Internet Things J., 1*(1), 22–32.

ZWIA, (2009). *Zero Waste Definition. Zero Waste International Alliance*. Available from: http://zwia.org/zero-waste-definition/ (accessed on 19 February 2021).

INDEX

Z